NATURAL RESOURCES OF SOUTH-EAST ASIA
General Editor: OOI JIN BEE

THE PETROLEUM RESOURCES
OF INDONESIA

THE PETROLEUM RESOURCES OF INDONESIA

OOI JIN BEE

Professor of Geography
National University of Singapore

Issued under the auspices of the
Institute of Southeast Asian Studies in Singapore

KUALA LUMPUR
OXFORD UNIVERSITY PRESS
OXFORD NEW YORK MELBOURNE
1982

Oxford University Press
Oxford London Glasgow
New York Toronto Melbourne Auckland
Kuala Lumpur Singapore Hong Kong Tokyo
Delhi Bombay Calcutta Madras Karachi
Nairobi Dar es Salaam Cape Town

and associates in

Beirut Berlin Ibadan Mexico City Nicosia

ISBN 978-94-011-7949-2   ISBN 978-94-011-7947-8 (eBook)
DOI 10.1007/978-94-011-7947-8

© *Oxford University Press 1982*

Softcover reprint of the hardcover 1st edition 1982

*Published by Oxford University Press, 3, Jalan 13/3,
Petaling Jaya, Selangor, Malaysia.*

For
Siew Khim and
Jin Eong

Preface

THE quadrupling of oil prices within a few months in late 1973 and early 1974 brought to an abrupt end the era of inexpensive oil. Since then the continuing increases in the price of oil traded in the international market and the higher prices of imports of manufactured goods have seriously disrupted the foreign exchange balances of many developing countries and forced them to replan their development programmes.

The impact of high oil prices is felt in every country, whether developed or developing, and has brought to world attention the fact that not only are petroleum resources in limited supply and exhaustible but also that substitutes cannot be found easily or quickly. In a world faced with the certainty of declining supplies of petroleum there is widespread interest and concern among all the oil producing countries to evaluate the extent of their petroleum resources and to examine more closely the problems of their development, rates of depletion and methods of conservation.

The present work reviews some of the above issues and problems in relation to Indonesia, an OPEC member, and the major oil producing country in South-East Asia. More specifically, it seeks to provide the reader with an overview of the petroleum resources of the country—their nature, extent, distribution as well as the problems of their development.

The study relies heavily on official data, both published and unpublished. Data gaps, where they exist, have been bridged by drawing upon diverse other sources of information such as regional and local journals, bulletins and newspapers.

I have discussed various aspects of my research with a number of key officials and scholars, in Indonesia and Singapore, and have benefited from their comments and suggestions. I would like in particular to acknowledge my debt to Dr Luki Witoelar Kartaadipoetra, Head, Geological Evaluation and Development, PERTAMINA, for providing me with unpublished material and for his assistance on the chapter on the petroleum reserves and resources of Indonesia. Mr Leslie R. Beddoes, Jr., General Manager, Cities Service East Asia, Inc. Singapore, evaluated the chapter on the geologic setting of Indonesia's petroleum deposits, clarified some points, and gave me the benefit of his geological expertise and experience of the Indonesian petroleum scene. I have also benefited greatly from discussions with the following officials in the PERTAMINA Head Office, Jakarta: Ir. Atik Suardy, Senior Geologist; Mr R. H. Subijanto, Head, Public Relations; J. A. F. Masiruw, Public Relations Department; Mr J. F. Menayang, General Affairs Manager; Mr Sukastoyo, Ka Humas dan Peremintahan, Badan Koordinasi Kontractor[2] Asing; and Mr D. L. Coutrier, Environment Protection Coordinator. Professor Dr Wahjudi Wisaksono, Director of Lemigas, Jakarta, was kind enough to spend time with me outlining some of the problems facing the petroleum sector in Indonesia. Ir. Soembarjono, Director of Exploration and Production, Directorate-General of Oil and Gas, Department of Mines and Energy, Jakarta, and Mr Fauzan Zailani Sh. of the same Department provided me with a great deal of useful information. My stay in Jakarta was made especially pleasant through the kindness of Mr Charles N. Silver, Cultural Office of USICA, American Embassy, who went out of his way to meet my requests for various documents and materials.

I would also like to express my thanks to Mr W. Weinhold of the USICA, American Embassy, Singapore, and to Mr Amhar Moelia, Head, and Mr Raymond Y. H. Lim, Administration—Finance Manager, Singapore Office of PERTAMINA.

The main body of this work was completed while I was a Visiting Fellow of the Institute of Southeast Asian Studies, on sabbatical leave from the National University of Singapore. I am grateful to the Institute for providing me with financial support for my field trip to Indonesia, and to the Director, Professor Kernial Singh Sandhu for his kind assistance at all times. My thanks should also go to the Librarian, Datin Patricia Lim and to her staff.

Mrs Lim Kim Leng and Mrs Irene Chee of the Department of Geography, National University of Singapore, cheerfully provided me with all the typing and secretarial assistance I needed, while Mr Poon Puay Kee, Senior Cartographer, was responsible for drawing the maps and diagrams.

I am, of course, solely responsible for the errors, omissions and shortcomings in this book.

Singapore   OOI JIN BEE
1981

GENERAL NOTE ABOUT STATISTICAL AND OTHER DATA

THE statistical data used in this book are drawn mainly from official sources.

The maps and diagrams were drawn by Mr Poon Puay Kee under the author's direction.

Values given in dollars refer to U.S. dollars.

Contents

Tables

Figures

I

The Development of the Oil Industry

The Pre-war Phase

INDONESIA was famous for its mineral wealth long before its spices attracted the attention of European traders and adventurers and drew them to its shores. An early Chinese reference to Java mentions the working of salt in AD 13, whilst archaeological and other evidence point towards indigenous familiarity with gold, copper, bronze and iron in the pre-Christian era. During the early colonial period tin was the main metal of commercial importance but the systematic exploitation of tin and other minerals, including oil, did not begin until the second half of the nineteenth century.

Oil from seepages in the Aceh region in Sumatra was used as fuel in naval battles along the Sumatra coast as early as the eighth century AD. Some old Dutch journals contain references to a 'wondrous substance' in the Malay Archipelago that 'is deemed inextinguishable once it has been ignited and that burns upon the sea'. It was also used for medicinal purposes by both the local inhabitants as well as the Dutch, especially for treating 'stiffness in the limbs'. The Dutch East India Company's envoys at the court of the Sultan of Acin were ordered to obtain quantities of this 'earth oil' for such medicinal use.

It was not until two centuries later that new and more important uses were discovered for this raw material. In the mid-nineteenth century a process was discovered for refining petroleum, or 'rock oil' as it was then known, into illuminating oil for lamps, and into lubricating oil, paraffin wax, and naphtha. This new illuminating oil soon displaced all other oils used for lamps. The great demand for this new oil stimulated a search for petroleum, initially in the United States, and later on extended to the other parts of the world, including

Indonesia (then the Netherlands East Indies). The Head of the Mines Department, C. de Groot, was confident that petroleum could be produced in commercial quantities in Indonesia, and as a first step in this direction, drew up a provisional list of the locations in which petroleum occurred. This list, completed in 1865, enumerated a total of fifty-two oil seepages and their yields. Additions were made to the list in the course of the next few years, and by 1870 a full picture of oil occurrences in Indonesia had been built up (Gerretson, 1953).

The first attempt at petroleum exploitation in Indonesia was made by Jan Reerink, a general store owner who decided to try his hand at this new venture. In 1871 he began drilling at Tjibodas (Jibodas), located at the foot of the volcano Tjarema (Jarema), south of Cheribon (Ceribon) in West Java. Although he found some high quality oil on two occasions, the yield was too small to be commercially viable, and after five years he had to give up his search for lack of capital.

In 1880 A. J. Zijlker, the Dutch manager of the East Sumatra Tobacco Company, discovered by chance the existence of oil pools in the vicinity of his plantation. He was able to secure a concession to the oil-bearing land (officially named Telaga Said) from the Sultan of Langkat, and in 1884 began drilling at Telaga Tiga, the most accessible of the oil pools at the concession. The first well gave some promising results but the second well drilled at Telaga Tunggal in 1885 was an immediate success, producing oil in commercial quantities at a depth of only 121 m.

In 1890 Zijlker made over his petroleum concession to a newly formed company Koninklijke Nederlandsche Maatschappij tot Exploitatie van Petroleumbronnen in Nederlandsch Indie (Royal Dutch Company for the working of petroleum wells in the Netherlands Indies) which later changed its name to the Royal Dutch Petroleum Company. In the first year of its operation the Royal Dutch produced 1 200 metric tons of oil from its Telaga Tunggal field. This well provided the raw material for the company's first refinery, established at Pangkalan Brandan in 1892.

The discovery of oil in commercial quantities at Telaga Said provided a great stimulus to exploration and drilling in other parts of Indonesia. New fields were discovered in Kruka, East Java (1887),

Kampong Minyak, South Sumatra (1896), Sumpal, South Sumatra (1897), Sanga Sanga, East Kalimantan (1897), Perlak, North Sumatra (1900) and Ledok, East Java (1901). All the fields had payzones located in the upper regressive sands lying at relatively shallow depths, usually with oil seeps in the neighbourhood providing surface indications of the oil at depth.

During this pioneer phase there were as many as eighteen different companies exploring and drilling for oil in various parts of Indonesia. Among them was the Shell Transport and Trading Company, founded by an Englishman, Marcus Samuel. The company originally traded in shells and spices but went into the oil shipping business and eventually also into oil exploration and drilling. In 1894 the company discovered oil in East Kalimantan and in the same year, established a small refinery in Balikpapan. However it failed to advance from this base in oil production, losing ground, as did the other small companies, to the fast expanding Royal Dutch. In 1907 the Shell Transport and Trading Company found relief from its difficulties by merging with the Royal Dutch on a 40/60 basis, the new company being named the Royal Dutch/Shell Group. Three operating companies were created under this parent holding company: the Bataafsche Petroleum Maatschappij (B.P.M.), handling production and refining; the Asiatic Petroleum Company, handling marketing; and the Anglo-Saxon Petroleum Company, handling transportation.

The Royal Dutch/Shell Group through its operating company, the B.P.M., gained full domination of the Indonesian oil industry within four years of its founding when it absorbed the last independent oil company, Dordsche Petroleum Maatschappij, in 1911. The group held 44 concessions totalling 32 000 sq. km, of which 19 were in Sumatra, 18 in Java and 7 in Kalimantan. The total Indonesian production in that year was 13 million barrels (3.4 per cent of world production), divided as follows:

East Kalimantan 34 per cent (main field: Mahakam River delta)
North Sumatra 22 per cent (main field: Perlak, Aceh)
South Sumatra 20 per cent (main field: near Muara Enim)
North-East Kalimantan 14 per cent (main field: Tarakan)
East Java 10 per cent (main field: near Tjepu (Jepu) and Surabaya).

The year 1912 was marked by the formation of a new oil company, Nederlandsche Koloniale Petroleum Maatschappij (N.K.P.M.), a subsidiary of the Standard Oil Company of New Jersey (later to become the Standard-Vacuum Company, and after the Second World War, P.T. Stanvac Indonesia). Its initial efforts were unproductive and disappointing, and by 1920 it had a production of only 100 barrels per day, against the Royal Dutch/Shell's production of 48,000 barrels per day (Bartlett, et al., 1972). However in 1919 N.K.P.M. obtained a seventy-five-year mining concession from the government of the Talang Akar area of South Sumatra, and after drilling some shallow wells, made a major discovery in 1922 when it found oil in the prolific lower transgressive sands, several hundred feet below any previously known producing horizon in Indonesia. Up to then all the payzones in Indonesia were in the upper regressive sands but after the Talang Akar discovery, augmented by a similar discovery in the adjacent Pendopo area in 1928, attention was focused on the bigger potential of the lower transgressive sands. All the other large oil fields—Minas and Duri in Central Sumatra, Limau in South Sumatra, and Tandjung (Tanjung) in South Kalimantan—that were subsequently discovered had payzones in these sands.

In 1921 the government decided to establish a new oil company, the Nederlandsche Indische Aardolie Maatschappij (N.I.A.M.), on a joint venture basis with B.P.M., the latter providing the operational management and marketing the products. The intention of the government was to redress the balance between the oil companies and the state, as it felt that the concession system operated too much in the favour of the companies. The new company obtained the exploitation rights of the oil basin in Djambi (Jambi), South Sumatra, a small area in North Sumatra, and the island of Bunyu in North-East Kalimantan. Of these Djambi proved to be the most important, producing over 60 million barrels between 1924 and 1940.

The oil industry in Indonesia expanded steadily and by 1924 the number of oil concessions had increased to 119, covering a total area of 6 400 sq. km. Total production had also increased to 22.6 million barrels, of which 95 per cent was from the Royal Dutch/Shell Group and the rest from the N.K.P.M. The relative importance of the regional output when compared with the situation in 1911 had also changed significantly: East Kalimantan remained the largest producer with 36 per cent, but Tarakan in North-East Kalimantan jumped to second place with 32 per cent, while production in North Sumatra declined to 6 per cent of the total Indonesian output. South Sumatra produced 17 per cent and Java 9 per cent.

The competitive position of the Indonesian oil industry in this early phase was also partly due to a Mines Act which allowed for the granting of concessions covering entire oil structures, thereby permitting each oil company to develop its fields as economically as possible. Concessions were granted for seventy-five years, with few obligations to drill being imposed on the companies. But in 1928 the government altered the regulations governing concessions, making them less favourable to the oil companies. Under the new regulations the concession period was reduced to forty years; moreover the oil company had an obligation to drill, with however, the right to return those parts of the concession area that had no oil prospects. Furthermore the company was obliged to pay the state royalties for the concessions as well as a progressive profit share amounting to as much as 20 per cent of the net profit.

In 1930 the Standard Oil Company of California formed a subsidiary company in Indonesia, Nederlandsche Pacific Petroleum Maatschappij (N.P.P.M.), but was not able to obtain oil concessions until 1936, when it was granted an area in Central Sumatra and another one in West Java. In that year Standard of California entered into a joint partnership with the Texas Corporation to form the California Texas Oil Company (Caltex). The company initiated extensive prospecting work on their concession areas, and in 1939 spudded its first exploration well in Central Sumatra. Commercial production began just before the outbreak of the Pacific War.

Up to the 1930s Irian Jaya, then known as Dutch New Guinea, remained outside the interest of all the oil companies except the Royal Dutch/Shell Group. In 1935 a new company was formed to undertake the systematic exploration of this large territory. Named the Nederlandsche Nieuw Guinee Petroleum Maatschappij (N.N.G.P.M.), this company was a joint interest of the Royal Dutch/Shell Group and the Standard-Vacuum Company (Stanvac) through N.K.P.M., each with a 40 per cent stake, and Caltex with a 20 per cent stake. The

TABLE 1.1

MAJOR OIL COMPANIES IN INDONESIA, 1940

Company	Area	Production in 1940 (million barrels)	Controlling Interest
N.V. de Bataaf-sche Petroleum Maatschappij (B.P.M.)	N. & S. Sumatra, E. Java, N.E. & E. Borneo, Ceram (Seram)	35.3 (57%)	Royal Dutch Group
N.V. Nederland-sch—Indische Aardolie Maatschappij (N.I.A.M.)	S. Sumatra (Jambi), Poelau Pandjang (N.E. Sumatra)	10.0 (17%)	N.I. Government and B.P.M.
N.V. Nederland-sch—Koloniale Petroleum Maatschappij (N.K.P.M.)	S. Sumatra (Palembang), E. Java (Rembang)	16.2 (26%)	Standard-Vacuum Co.
N.V. Nederland-sche Pacific Petroleum Maatschappij (N.P.P.M.)	Mid. Sumatra (E. Coast), W. Java	Not known	Standard Oil Co. of California and Texas Corporation
N.V. Nederland-sche Nieuw Guinee Petroleum Maatschappij (N.N.G.P.M.)	New Guinea (Vogelkop)	None	40% Royal Dutch Group; 40% Standard-Vacuum through N.K.P.M; 20% Caltex through N.P.P.M.

company secured a fifty-year concession in the western part of Irian Jaya, but its initial exploration efforts were greatly hampered by the area's isolation, its difficult terrain, and by labour-recruitment problems. However it did manage to carry out intensive exploration over an area of 100 000 sq. km. Wells were drilled in the Vogelkop, at Klamono, Wasian and Mogoi. Although some promising results were obtained from these exploration wells, the war prevented further development of this part of Indonesia.

This early phase in the history of the petroleum industry of Indonesia is summed up in Table 1.1, listing the five major oil companies and their production in 1940. Production in that year amounted to 61.5 million barrels. The daily production rate was 136,000 barrels. Over 1 million metric tons of gas were also produced from the fields. The main oil company, B.P.M. of the Royal Dutch/ Shell Group, had by this time built six refineries, two in each of the main producing areas of Sumatra, Java and Kalimantan, while the N.K.P.M. had a refinery in Sumatra and another one in Java. The crude oil from the N.I.A.M. fields was treated in the B.P.M. refineries.

About 31 million barrels of oil were exported in 1940, much of it through Singapore by way of the entrepôt installations on the Indonesian offshore islands, as the Sumatran oil ports of Pangkalansusu and Palembang were unable to load ocean-going tankers to full capacity because of shifting sandbars.

The War Years

One of the prime targets of the Japanese invasion of South-East Asia in the Second World War was resource-rich Indonesia, particularly its petroleum which the Japanese needed to drive its war machine. The Dutch colonial government realized that they would be unable to defend their colony against the Japanese, and consequently sought to deny them the oil facilities by following a scorched-earth policy. Refineries were burned, wells plugged with cement, and pipelines dynamited, but because of the hurried nature of the job and the swiftness of the Japanese invasion, some facilities were left un-touched.

The expected invasion took place on 12 January 1942 when the Japanese landed at Tarakan, just after the Dutch army had set fire to the oil installations. By 23 January the Japanese had advanced to Balikpapan. Another attack was mounted in Sumatra, with Palembang falling on 15 February. By the end of March the Japanese army was in total control of all the oil fields in Indonesia.

The immediate task of the Japanese was to restore production from the oil fields, but their rehabilitation efforts were necessarily makeshift and hindered by inexperience, materials and equipment shortage and lack of skilled labour. Nevertheless they did manage to get the oil flowing from the wells, particularly those that had been abandoned and had not been plugged. Production figures for the war years are not reliable, but it is estimated that about 24 million barrels were produced in 1942, 48 million barrels in 1943, 24 million barrels in 1944 and only 7.6 million barrels in 1945 when Allied bombing attacks destroyed many of the oil installations (Bartlett, *et al.*, 1972).

The Early Post-war Period (1945–1966)

The Japanese surrendered on 15 August 1945. Two days later Indonesian leaders, determined to free themselves from Dutch colonial rule, proclaimed independence for their country. The period between then and December 1949 when independence was finally attained was one of political instability as the Indonesians and the Dutch fought on the battlefield and at the negotiating table for the control of the islands.

Under such circumstances it was not surprising that the oil industry made slow progress towards recovery. The Indonesian leaders had embodied rights to the country's natural resources in Article 33 of the 1945 Constitution which stated that all resources of the land and water of Indonesia belonged unequivocally to the people. In the interim period between the defeat of Japan and the arrival of the Allied occupation forces the Indonesians took over the oil fields and refineries as *de facto* owners. The return of the Dutch and their refusal to relinquish power to the Indonesians resulted in a struggle for control which affected the oil industry as much as it did the other sections of the economy.

B.P.M. was able to resume production in Tarakan and the Kalimantan mainland between 1945 and 1946, but neither B.P.M. nor N.K.P.M. could enter their former oil fields in South Sumatra until 1947. In North Sumatra the Japanese Occupation Army transferred control of the oil fields in 1945 to the representative of the Indonesian oil workers. An Indonesian oil company, the first to be formed, was established under the name Perusahaan Tambang Minyak Negara Republik Indonesia (The Indonesian Republic Oil Company) to work the fields in the Pangkalan Brandan area and the Aceh region. The former B.P.M. oil fields in the Kawengan area of Central Java were similarly taken over by an Indonesian oil workers' co-operative, and eventually came under the control of the state enterprise PERMIGAN (Perusahaan Minyak dan Gas Nasional —The National Oil and Gas Company). In Irian Jaya (New Guinea), isolated from the political unrest of Java and Sumatra, oil exploration activities were resumed in 1946, with the Klamono field in the Vogelkop producing 4,000 barrels per day in 1948. Under the troubled conditions of these early post-war years production of crude oil in Indonesia could only increase slowly: from 2.3 million barrels immediately after the war in 1946 to 46 million barrels in 1949, still well below the 1940 production of 61.5 million barrels. It was only in 1951 that production, at 63 million barrels, reached the pre-war level.

In the early post-war period both Shell and Stanvac had tried to obtain new exploration and development areas so as to maintain their output, but without success. Soon after independence Indonesia decided to postpone the granting of concession and exploitation permits. Its attitude hardened in 1957 when Parliament passed a resolution stopping new leases from being granted. These two major oil companies were therefore solely dependent on their existing (pre-war) concessions. Most of the onshore Indonesian fields were small, requiring a large number of wells to extract the oil. The output of Shell was from 18 fields with over 1,000 wells, while Stanvac's fields in Central and South Sumatra required some 500 wells to produce its oil. Moreover, a high proportion of both companies' wells were on pump. In these circumstances output from the fields would soon decline unless boosted by a continuous extension of the area under production.

Stanvac had about 10 per cent of all long-term pre-war concession areas in Indonesia, amounting to 728 000 hectares. Most of this was in South and Central Sumatra, a region already extensively developed and thought to hold little further potentialities. Shell in the 1950s had a larger concession area than Stanvac, but part of the area was shared with N.I.A.M. This part was eventually handed over to

PERMINDO (Pertambangan Minyak Nasional Indonesia—Indonesian National Oil Company), the state enterprise which took over from N.I.A.M. in 1959. Moreover Shell could not regain access to the Central Java or North Sumatra fields, taken over by Indonesian oil workers in 1945. Its other concession areas in Kalimantan and East Java were close to being worked out.

In contrast to these pre-war oil companies, the Caltex Pacific Oil Company, which began its exploratory programme only in 1936, and whose activities were interrupted by the war, had most of its concession areas still untapped. It was therefore able to launch an intensive search for oil soon after the war, and made the most productive oil strike in Indonesian history when it discovered the giant Minas field in Central Sumatra, estimated to be capable of producing several billion barrels of oil. This field, together with the important Duri field which came into production just before the war, enabled Caltex to dominate the post-war production scene, to the extent that in 1957 it was responsible for nearly half of the total Indonesian production of crude, and by 1963 for more than half. In contrast the oil companies of Shell, N.I.A.M. and Stanvac, the main pre-war producers, slipped from their top positions to occupy relatively minor places in this post-war phase (Table 1.2).

The output of crude oil in Indonesia in the period under survey rose substantially—from 63 million barrels in 1951 to 177 million barrels

TABLE 1.2

PERCENTAGE PRODUCTION OF INDONESIAN
CRUDE OIL, 1925–1963

Company	Year			
	1925	1940	1957	1963
Shell	95	57	23	26
Stanvac	5	26	20	11
N.I.A.M.	—	17	10	3
Caltex	—	—	47	55
PERMINA (state oil enterprise)	—	—	—	5
Total	100	100	100	100

in 1965 (280 per cent). Most of this increase was from newly-exploited fields discovered just before the war and brought into production in the post-war period. As noted earlier the most important of these fields was the Minas field in Central Sumatra. Together with a much smaller output from the Duri field, the Minas field enabled Caltex to dominate oil production in Indonesia from the mid-1950s onwards.

In contrast the older pre-war fields of Shell and Stanvac had reached their production peak before the war and were on the decline soon after the war. Shell's output dropped in the mid-1950s but the development of a new field at Tanjung enabled it to recover its output to the level attained in 1950—4. In relative terms, however, Shell did not do well, its contribution to the total Indonesian output increasing by only 3 per cent between 1957 and 1963 (Table 1.2). Stanvac similarly had to face the problem of falling production from old fields, mainly in South Sumatra. Its share of the Indonesian output of oil fell from 20 per cent in 1957 to only 11 per cent in 1963 (Table 1.2).

The Post-1966 Period

The period 1965—6 could be regarded as a watershed in the history of the Indonesian oil industry. The Indonesian Communist Party's abortive coup of October 1965 brought about a wave of repercussions which eventually affected all aspects of Indonesian life and economy, including the oil industry. Initially there was a period of uncertainty when the tide of confidence dropped and investment in the oil sector came to a virtual standstill. The uncertainties of these months were reflected in the output of crude, which fell from 177 million barrels in 1965 to 170 million barrels in 1966. In this year oil contributed only 5 per cent to the country's revenue. In fact the period between the passing of the Petroleum Law in 1960 and the end of the Sukarno regime in 1967 was one of stagnation for the oil industry in Indonesia, with production increasing at an average rate of 2.2 per cent per annum, most of this increase being from the Tanjung field which started production in 1961.

However, as the Suharto regime settled down to its work of charting a new course for Indonesia and demonstrated its determination to correct the economic transientness of the Old Order, the tide of confidence was restored, and output climbed to higher levels in subsequent years as new fields came into production.

The year 1966 also witnessed a shift in the focus of interest among the oil companies—from onshore to offshore oil. This was mainly due to advances in technology which made it possible to drill for oil in offshore areas. The geological indications were that Indonesia's extensive continental shelves held good prospects for oil. Furthermore, drilling conditions were very favourable, the seas being shallow and calm and the temperatures warm throughout the year. It has been estimated that for every fifty days lost time experienced in the North Sea, only twenty days are lost in South-East Asia. Moreover, the oil companies would save on production costs as low-cost production platforms could be used in the warm shallow seas of Indonesia (Adams, 1980). The search for offshore oil, however, did not actually begin until after the foreign oil companies had accepted the principle of production-sharing and signed production-sharing contracts whereby the host government of Indonesia and the foreign oil companies shared the oil produced and not the profits realized.[1] After an initial period of uncertainty, thirteen oil companies signed production-sharing agreements between 1966–8, ushering a new era in the history of the petroleum industry of Indonesia.

These companies mounted an intensive search for oil, concentrating their efforts on offshore areas. Sinclair Exploration Company, which later merged with Atlantic Richfield (ARCO), started drilling in the Arjuna Complex in the Java Sea in September 1968, and struck oil in February 1969 at a depth of 1 400 m. This, the first offshore oil to be discovered, was a low-sulphur crude which tested at a flow rate of 2,600 barrels per day.

Kyushu also initiated drilling towards the end of 1968, offshore of Banjarmasin in South Kalimantan, but its efforts were unsuccessful and it abandoned the search in late 1969. Other companies were more successful—Union Oil of Indonesia discovered a field in the Attaka area, off East Kalimantan, while IIAPCO (Independent Indonesian American Oil Company) found offshore oil in the Java Sea (the Cinta

---

[1] The contractual arrangements under which the oil companies operate are discussed in greater detail in the next chapter.

field) and south-east of Sumatra (the Kitty field) with estimated capacities varying from 3,000 to 7,500 barrels per day. These finds were made in 1970, and provided a stimulus to other oil companies to sign new production-sharing agreements, and to those who had already done so to step up their exploration activities. By the end of 1970 the newly-formed state oil company PERTAMINA had concluded thirty production-sharing agreements with oil companies, and by 1974 virtually all of Indonesia's offshore areas had been contracted out.

Commercial production of offshore oil started in late 1970, in the Cinta and Arjuna (North-West Java) offshore fields. The average daily production in 1971 from these two fields was 6,760 barrels and 4,140 barrels respectively. By 1974 offshore crude oil production from these and other newly-discovered offshore fields had increased to 18 per cent of the total Indonesian production. In contrast to the 1960–6 period when production stagnated, the post-1966 period up to 1973 witnessed a dramatic increase in total output, with the annual increase averaging 16 per cent.

Although offshore oil has made a progressively important contribution to Indonesian oil output in the 1970s, the bulk of such output continued to be from onshore sources. By far the biggest onshore producer was Caltex, whose Minas field in Central Sumatra alone contributed an average of 72 per cent of the total Indonesian output during the period 1970–3. Caltex has a 60/40 profit-sharing Contract of Work under which it has been operating since 1963 and which is due to expire in 1983. The history of Caltex's subsequent (i.e. post-1963) working arrangements with the national oil company PERTAMINA illustrates the flexible attitude which the latter, under its former President-Director, Dr Ibnu Sutowo, adopted in seeking to maximize its earnings from the country's oil resources. Thus in 1968 when Caltex applied for a 4,144 sq. km extension of its contract area in Central Sumatra under the old terms, PERTAMINA agreed to the request because Dr Ibnu Sutowo was convinced that Indonesia would gain more profits by letting an established company with lower development costs operate in the area under the old Contract of Work rather than by letting it to a new company under production-sharing terms. Similarly, in pursuance of

TABLE 1.3
DISCOVERY OF OIL FIELDS, BY DECADE

Decade	Number of Fields Discovered
1890–1899	3
1900–1909	2
1910–1919	2
1920–1929	11
1930–1939	13
1940–1949	11
1950–1959	12
1960–1969	14
Total 1890–1969	68
1970	11
1971	14
1972	20
1973	24
1974	29
1975	20
1976	22
Total 1970–6	140

Source: A profile of Indonesia's petroleum industry, 1979.

the maximization objective, PERTAMINA in 1971 extended the operating life of Caltex's existing 971 000 ha in Central Sumatra from 1983 to the year 2001 under a production-sharing contract, and signed a new production-sharing contract with the company for an additional 2 226 000 ha of land adjacent to the existing holdings (Knowles, 1973).

The 1970s have seen the discovery and production of oil on a scale dwarfing that of all previous decades. One indication is that the number of oil fields discovered in the seven years 1970–6 was ten times the number discovered in the decade 1960–9, and more than twice the total number discovered in the eighty years 1890–1969 (Table 1.3). The new fields encompass the span of the Indonesian archipelago, including the major islands as well as the continental shelves. Production increased steadily from 170 million barrels in 1966 to a high of 615 million barrels in 1977, an average annual increase of 13 per cent over eleven years. However production fell slightly in 1978 and again in 1979, to 598 million barrels and 583 million barrels respectively (Fig. 1.1). The 1979 production figure was about 20 per cent below the 730 million barrels projected under the Indonesian Second Five-Year Plan for that year.

The decline in production is a consequence of the fall in output from many of the old fields and the exhaustion of some. A characteristic feature of the petroleum geology of Indonesia is that, unlike the Middle Eastern fields, most of the fields are small and new fields will have to be discovered and brought into production to sustain output. The generally small size of the oil fields is a result of the fact that the Tertiary basin architecture in Indonesia (and other

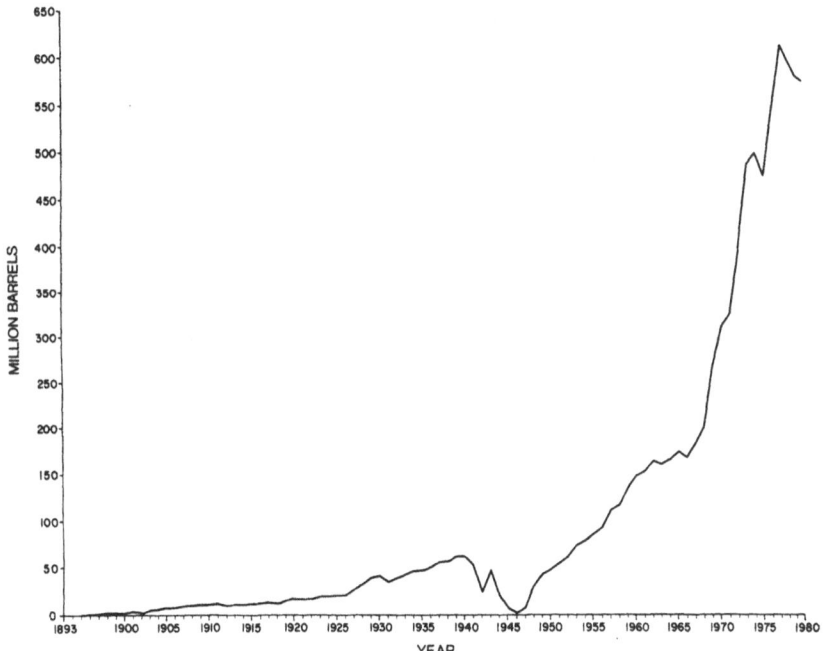

Fig. 1.1   Crude oil production in Indonesia, 1895–1980
Sources: 1893–1971: OPEC Annual Statistical Bulletin
1972–1980: PERTAMINA.

TABLE 1.4
WELL PRODUCTIVITY IN INDONESIA, 1972–1979

Year	Total Daily Production (Barrels/Day)	Number of Producing Wells (as of 1 July)	Average Output Per Well Barrels/Day
1972	1,027,000	2,344	438.1
1973	1,300,000	2,567	506.4
1974	1,457,000	2,710	537.6
1975	1,300,000	3,018	430.8
1976	1,500,000	3,162	474.4
1977	1,690,000	3,421	494.0
1978	1,650,000	3,644	452.8
1979	1,600,000	3,823	418.5

Source: Compiled from data in the end-of-year issues of the *Oil and Gas Journal*.

parts of South-East Asia) does not provide the trapping mechanisms for major oil fields (Klemme, 1975). Except for a few fields in the Central Sumatra basin (notably Minas) and the Kutei basin (Handil) the average size of the oil fields is less than 20 million barrels of recoverable oil. Most wells produce less than 750 barrels per day, and field productivity is correspondingly low, being less than 30,000 barrels per day (Adams, 1980). In fact data compiled from the year-end issues of the *Oil and Gas Journal* show that the average output peaked at 537.6 barrels per day per well in 1974 but has since declined to only 418.5 barrels per day in 1979. The total number of producing wells increased by 41 per cent between 1974 and 1979 but because of the decline in the average productivity of the wells total daily output increased by only 10 per cent (Table 1.4).

In 1977 when Indonesian crude oil production peaked at a total of 615 million barrels nearly half of the total production was derived from fields which were expected to be exhausted within the next three to four years. Consequently unless new fields were discovered within this period production could be expected to fall, as in fact it did in 1978 and again in 1979, by over 2 per cent in each of these years.

The fall in production had its roots not only in the rapid decline rates of the characteristically small reservoirs of Indonesia (which were not compensated for by any significantly large oil finds) but also in the low level of exploration conducted by the foreign oil companies in the years immediately following the PERTAMINA financial crisis of mid-1975. The crisis, which is discussed at greater length in the next chapter, saddled the Indonesian government with a debt of massive proportions, and forced it to try and raise additional revenues by the renegotiations of its contracts with the oil companies. The outcome of these renegotiations took the form of new contractual agreements which benefited the state by increasing the governmental 'take' of oil produced, while necessarily reducing the oil companies' net revenues. The new terms were regarded by the companies as a serious disincentive to exploration, and the level of exploration fell sharply in the years 1976 and 1977.

The drop in crude oil production and consequent fall in foreign exchange earnings was partially compensated for by the contribution, from 1977 onwards, of Indonesia's other major hydrocarbon resource—natural gas, exported in the form of liquefied natural gas (LNG). The contribution of LNG to the country's foreign exchange earnings has been significant, and will grow substantially in the years to come.

For a number of reasons which will be examined in Chapter 4 the level of exploration activities increased in the years 1978 and 1979, thus reversing the downward trend of the previous two years. The upturn in exploration may in due course arrest the decline in production of 1978−9. The Third Five-Year Plan forecasted small falls in output in 1979 and 1980, followed by increases from 1981 onwards so that by 1983−4 production would reach 1.83 million barrels per day (Table 1.5).

Exports of crude oil from Indonesia reached a peak of 484 million barrels in 1977, from a record total production of 615 million barrels. Since then there has been a downward trend, partly because of reduced production, but also as a result of increases in the domestic demand for oil. By 1980 the volume of oil exported had dropped to 390 million barrels, or 81 per cent of exports for 1977.

The main export destinations in the 1970s were Japan and the U.S. which together bought between 79 and 93 per cent of the oil from Indonesia. In 1979 Japan purchased 57 per cent of the total exports of crude oil, including condensate, while the U.S. imported 29.5 per

TABLE 1.5
PROJECTED OIL PRODUCTION UNDER REPELITA III
(Third Five-Year Plan)

Fiscal Year	Annual Production (Million Barrels)	Average Per Day (Million Barrels)
1979−80	582	1.57
1980−1	572	1.56
1981−2	604	1.65
1982−3	640	1.75
1983−4	668	1.83

cent. The other export destinations were Trinidad (6.2 per cent) and the Philippines (2 per cent). The remainder of the oil went to a number of other countries. It appears likely that Japan and the U.S. will maintain their positions as the leading customers of Indonesian crude, being able and willing to pay a premium price for the low-sulphur oil produced in Indonesia. Japan at present is also the export destination of Indonesian LNG, with the U.S. likely to be the second one as soon as outstanding problems, including the location of the receiving terminal in California, are resolved.

II

PERTAMINA: The National Oil Company

The Early Years

ONE of the main provisions of the 1945 Indonesian Constitution is Article 33 which states, *inter alia*, that:

Branches of production which are important to the State and which affect the life of most people, shall be controlled by the State (para 2).

Land and water and the natural resources contained therein shall be controlled by the State and utilized for the greatest welfare of the people (para 3).

In so far as mineral resources, including petroleum, are concerned the concept of the state's sovereignty over such resources was not a novel one as the 1899 Netherlands East Indies Mining Act (Indische Mijnwet) had also laid down that rights over land (surface rights) conferred no similar rights over the mineral resources, these being vested in the government.

However, the old Mining Act allowed for the granting of mining concessions by the Governor-General to Dutch subjects and companies registered in the Netherlands or the Netherlands East Indies. The concession-holder thereby acquired the rights to mine the mineral resources in an area of land defined in the concession and hence a direct right over these resources.

This was patently contrary to the intent of Article 33 of the Indonesian Constitution, which vests such rights exclusively to the state. Accordingly, soon after Indonesia attained its independence the government established a State Commission on Mining Affairs to look into the operations of all mineral rights in Indonesia as well as to draft a new mining law. Pending the report of the Commission, the Indonesian Parliament prohibited the granting of new concessions.

The Commission submitted its report in 1959 and in 1960 the Netherlands East Indies statute and ordinance were replaced by a new Indonesian Mining Law (Law No. 37/1960). A new oil and gas law (Law No. 44/1960) was also promulgated to cover oil and natural gas specifically. The main provisions of Law No. 44/1960 were:

1. All oil and natural gas are national assets controlled by the state and its extraction can only be undertaken by state enterprises.

2. All mining rights are granted only to state enterprises.

3. To the Minister is given the right to appoint another party as contractor of the state enterprise for so far it is unable to develop the oil industry on its own, due to lack of capital and skill.

The new law not only abolished the old concession system but re-affirmed the principle of national sovereignty over the oil and gas resources, and strengthened this by stipulating that only the state or a state enterprise is authorized to exploit such resources. However, aware that the development of the oil and gas resources called for large investments of capital, advanced technology, experience and skills, the new law accordingly provided for the participation of foreign oil companies, not as concession-holders, but as contractors to the holders of the mining rights, that is, the state or a state enterprise. The three foreign oil companies operating in Indonesia—Caltex, Stanvac and Shell—were asked to adjust their operations to conform with the terms and conditions of this new regulation.

This in effect would involve a major change in the working relationships between the oil companies and the government, a change that the oil companies were unwilling to accept. A prolonged period of negotiations between the two parties followed, with the government insisting on changing the status of the oil companies from concession-holders with ownership rights in their concessions to contractors to the state, and on a 60 per cent share of the oil profits.

In 1962 while the negotiations were taking place, an oil company new to the scene, Pan American Oil, concluded an agreement with the Indonesian government on terms which were a major departure from the old concession agreements and which conformed with the intent of the new law. The agreement was for a thirty-year period, with an exploration commitment spread over eight years covering 3.64 million ha in Central Sumatra. It included a cash bonus of $5 million (which may be regarded as a form of ground rent) and provided for other cash bonuses when production reached certain levels, and for all profits to be shared on a 60/40 basis. This agreement was to serve as a model for the later agreements with the three major oil companies operating in Indonesia.

In 1963, following an ultimatum issued by the Indonesian government to the three oil companies an agreement was reached and signed in Tokyo on 1 June. The Tokyo Agreement, as it was known, had five major sections (Hunter, 1965):

1. Each of the three foreign enterprises gave up its concession ownership rights granted under the old N.E.I. government and agreed instead to act as contractor to one of the three state oil companies.

2. In exchange they were awarded 20-year contracts to continue the development of the old concession areas: and were also permitted to make application for 30-year contracts to explore and develop new ones. The new area contracts required immediate payment of cash bonuses of $5 million, and continued tenure was subject to certain conditions.

3. Marketing and distribution facilities were to be handed over to the counterpart state companies within *five years* at prices based on agreed formulae. The foreign companies agreed to supply products to the state distribution organization at cost plus a fee of U.S. 10 cents per barrel for as long as required; pending transfer, distribution would be performed by the companies, for an additional fee of 10 cents per barrel.

4. Refining assets would be handed over within *10–15 years*, again according to agreed formulae; subsequently the companies would be prepared to supply crude oil to state refineries at cost plus a fee of 20 U.S. cents per barrel for as long as required.

5. The operating profits of the international companies from 1 June 1963, would be divided in the ratio 60/40 per cent between government and company—but in all events the government would receive a minimum payment of 20 per cent of gross value of crude oil produced in any one year.

The agreement further provided for the periodic relinquishment of portions of each new area to ensure that the company would actively explore its contract area. At the end of five years the company would have to relinquish 25 per cent of the new area, and a further 25 per cent at the end of ten years. The relinquished areas should be of a size and shape which would allow for further exploration, either by another company or by the state enterprise. All permanent installations in the relinquished area would become the property of the state enterprise.

The final agreements were signed on 25 September 1963, the details of the agreements being incorporated in contracts which became known as Perjanjian Karya or Kontrak Karya (Contract of Work). Although they no longer enjoyed the freedom that went with being concessionaries, the oil companies found the change in status acceptable as management was still in their hands. In reality they continued to exercise considerable freedom in their operations, with the government acting only in a supervisory capacity. As Dr Kusumaatmadja[1] has observed: 'The hard economic fact was that both sides could not really afford the petroleum industry to grind to a halt' (Kusumaatmadja, 1974).

It was logical that after the enactment of Law 44 of 1960, which in effect signalled the end of the old oil era in Indonesia, the government should itself participate more fully in the activities of the oil industry. The vehicle for such participation already existed in the form of the three national oil companies—PERMINA, PERTAMIN and PERMIGAN. All three companies were in operation when the new law came into effect. PERMINA (Perusahaan Minyak Nasional Indonesia—National Indonesian Oil Company), a limited liability company, was formed in 1957 when Dr Ibnu Sutowo was empowered with the task of establishing a company to undertake the rehabilitation of the oil industry and the export of oil to foreign countries. With the passing of the new law in 1960 PERMINA was converted into a state-owned enterprise and ceased to be a limited liability company. As part of the policy of running the oil and gas industry on a monopoly basis by the state, the Indonesian government through PERMINA gradually took over some of the assets and activities of the foreign oil companies. In 1964 B.P.M. handed over the entire assets of N.N.G.P.M. in Sarong, Irian Jaya to PERMINA. In 1965 PERMINA bought over the assets of Shell, including the refineries at Plaju and Balikpapan and the oil fields at Limau and Tanjung, as well as the marketing assets of Shell, Stanvac and Caltex. Its role was to take charge of oil exploration, production and marketing to foreign countries.

Negotiations on the take-over of Dutch-owned firms in 1958

[1]Dr Kusumaatmadja is now the Minister of Foreign Affairs, Indonesia.

resulted in the transfer of N.I.A.M to a newly-created limited liability company, PERMINDO, with Shell retaining 50 per cent ownership as well as providing the technical assistance. However under the new mining law PERMINDO was dissolved, and its place was taken by Pertambangan Minyak Indonesia (PERTAMIN), a state enterprise. PERTAMIN initially undertook some exploration work in areas first explored by N.I.A.M as well as the rehabilitation of several wells in Jambi. But these efforts were on a modest scale as the company funds were limited and it did not seek foreign capital for its exploration and development programmes.

Part of its operations was directed towards the marketing of oil products, and it was in this area that PERTAMIN eventually concentrated most of its efforts. Initially the company was assigned the task of supplying and distributing oil products only to the domestic market and the Armed Forces. In 1966 the activities of the two main national oil companies were rationalized when PERMINA was entrusted with the production functions of the oil industry and PERTAMIN took over the marketing of all fuel and lubricants.

The third national oil company PERMIGAN was established in 1961. It was the smallest and most poorly endowed of all the companies, having as its main assets some old oil fields in Central Java producing only 800 barrels of crude oil a day. PERMIGAN subsequently acquired two small fields—Kawengan and Wonosari—thereby boosting its total productive capacity to a peak of 2,700 barrels a day in 1963. However the company was unable to increase or even sustain this output through bringing new areas into production. The gradual exhaustion of its old fields led to a steady decline in output so that by 1965 only 1,800 barrels were being produced a day.

PERMIGAN's existence came to an end soon after the abortive coup of October 1965, when the board of directors was dissolved and its marketing facilities were handed over to PERTAMIN while its production responsibilities in West Java and Seram were assigned to PERMINA.

Besides the three state oil companies, the first half of the 1960s also saw the establishment of the national Petroleum and Natural Gas Institute (Lembaga Minyak dan Gas Bumi—LEMIGAS), an institute concerned with research on oil-related matters and which also ran an academy for the training of oil technicians. In addition, LEMIGAS also operated a few small oil fields in Java.

The period 1963−5 was a difficult one for the foreign oil companies as they had to face a series of wildcat strikes and take-over problems from increasingly militant communist-dominated labour unions. Shell, in consequence, drew up plans to withdraw all its Indonesian operations, and soon after the abortive October 1965 coup transferred all its assets to the Ministry of Oil and Gas Affairs for the sum of $110 million, to be paid in cash and in kind (in the form of crude oil). Stanvac had also offered to sell to the government its Sungei Gerong refinery and its assets in adjacent fields in South Sumatra, but negotiations were not carried through to a conclusive stage.

Further major changes were to follow. In 1966, General Suharto appointed the President-Director of PERMINA, Dr Ibnu Sutowo, to be Minister of Mines and Minister of Oil and Gas. Dr Ibnu Sutowo had been critical of the 1963 Contracts of Work, regarding them as being no more than 'dressed-up concession agreements'. As he put it:

The most important thing in the difference between a concession and a non-concession is the matter of ownership of the oil. The fundamental principle underlying a concession, and highly favoured by foreign oil companies, is that oil is owned by the government in its natural geological form, but as soon as man has done something to it, he is the owner of the oil. In other words, oil at the well-head becomes foreign property and the company pays a tax in the form of 60 per cent of the profits from the sale of the oil. Under the work contracts, whenever we want oil we have to request it from the foreign oil company, and we can ask for and obtain a maximum of only 20 per cent of gross production. These two aspects indicate that the work contract is no better than the old concession, and this should be opposed. Furthermore, if we are to adhere to the principle that the oil belongs to the Indonesian people, then, first of all, the management of the oil fields must be placed in Indonesian hands.

(quoted in Knowles, 1973.)

Dr Ibnu Sutowo was of the view that the solution to this problem was to establish a production-sharing system whereby the two parties involved—the host government and the foreign oil company—shared the oil produced and not the profits realized. The host government should also have management control. One of the advantages of the production-sharing system would be to remove all suspicion as to whether the price of oil used to calculate the profit split, in the profit-sharing system, was fairly arrived at.

However the major oil companies were not ready to accept the production-sharing concept enunciated by Dr Ibnu Sutowo, nor were they willing to relinquish management control as a condition for offshore oil exploration and production.

In such circumstances it was left to the small independent oil companies, who were generally more flexible in their approach, to step into the breach. The first of such companies on the scene was the Independent Indonesian American Oil Company (IIAPCO). This small group of American oil men had tried unsuccessfully in 1964 and 1965 to obtain a contract to search the area north-west of Java for offshore oil. They returned to Indonesia in early 1966 and after two months of negotiations, signed a production-sharing agreement with PERMINA, the first of such agreements in the international oil industry's history.

The agreement provided for the state enterprise to have management control, although, as explained by Dr Ibnu Sutowo, the spirit of the contract would be one of partnership, with the state enterprise the senior partner. IIAPCO would bear the pre-production risks, and if oil was discovered, its exploration cost recovery would be limited annually to 40 per cent of the oil produced. The remaining 60 per cent would be shared on a 65/35 basis, the larger share going to the state enterprise. All equipment brought into Indonesia by IIAPCO would become the property of the state enterprise, with the cost recovered under the 40 per cent formula. The contract area covered 54 776 sq. km offshore of North-West Java. IIAPCO would spend US$7.5 million over a period of six years in exploring the contract area. The contract itself was for thirty years. Between August 1966, when he signed the IIAPCO agreement, and November 1966, Dr Ibnu Sutowo concluded similar production-sharing agreements with four other small oil companies—Japan Petroleum Exploration Company (JAPEX), Refining Associates Canada Ltd. (REFICAN), Kyushu Oil Development Company Ltd., and Asamera Oil Indonesia Ltd.

In July 1966 Dr Ibnu Sutowo had stepped down as Minister of Mines, preferring to retain his position as President-Director of

PERMINA, and Director-General of the Bureau of Oil and Gas (Biro Minyak dan Gas—MIGAS). The man who succeeded him, Ir. Bratanata, held different views on the manner Indonesia's oil resources should be developed for the long-term benefit of the country. He was of the opinion that profit-sharing from the sale of oil, as provided for in the earlier Contracts of Work, was an adequate vehicle for obtaining just returns from Indonesian oil. He similarly held that the management provisions in the Contracts of Work should be retained. The way to maximize returns, he argued, was to divide the oil provinces into small blocks to ensure rapid exploration and development by a large number of companies. These blocks should be put up for competitive bidding, with the successful companies paying cash bonuses on the signing of contracts. His views were backed by the major oil companies who did not wish to see profit-sharing being replaced by production-sharing, or management control passing to Indonesian hands. They were naturally apprehensive as to the precedential effect of production-sharing on their interests in other parts of the world.

The divergent proposals of Dr Ibnu Sutowo and Ir. Bratanata were submitted to the Presidium Cabinet in January 1967. The Cabinet decided in favour of the production-sharing contracts, and by so doing, altered irrevocably the former patterns of relationships between the host government and foreign oil companies. The Cabinet also severed the Bureau of Oil and Gas from the Ministry of Mines and placed it directly under the Cabinet, thus clearing the way for Dr Ibnu Sutowo to follow his policies without referring to Ir. Bratanata's Ministry of Mines.

The terms contained in the IIAPCO's agreement formed the model for subsequent production-sharing contracts. Unlike the earlier Contracts of Work the IIAPCO agreement was between the state enterprise, PERMINA, and the oil company, and not between the government and the oil company. In essence the production-sharing contracts developed in 1966 embodied the following features:

1. The production-sharing contract, as its name implies, calls for the sharing of production (of crude oil) rather than the profits accrued from the sale of such oil.

2. The state or the state enterprise, as the owner of the petroleum

resource, retains control over the management of operations. (In practice, it is the contractor who executes the work programme. However, the contractor is required to submit an annual work programme and a budget 'setting forth the petroleum operations which the contractor proposes to carry out during the ensuing contract year'. The state enterprise has the right to propose revisions to these.)

3. The contractor is to provide all the financing for the operation and to bear all the pre-production risks. Where oil is discovered and produced the contractor can recover these costs up to a maximum of 40 per cent of the annual production.

4. Of the remaining 60 per cent of the annual production, the contractor is entitled to 35 per cent, and the state enterprise to 65 per cent.

5. The contractor is solely responsible for furnishing all the foreign exchange, materials, equipment and supplies necessary for the operation. Title to all project-related equipment brought in by the contractor would pass to the state enterprise upon entry into Indonesia. However, the contractor can recover the cost of this from the 40 per cent of output set aside for cost recovery.

Once the principle of production-sharing was established and officially endorsed, the oil companies (except those operating under Contracts of Work) had no choice but to accept it if they wanted to enter the Indonesian oil arena. Most companies attempted to modify or have deleted the management control clause, but all failed. Dr Ibnu Sutowo, as an inducement to the companies to sign contracts, promised that the first comers would receive more favourable terms and a pick of location. Late comers would have to sign contracts which would become progressively more favourable to the government and less so to the companies.

In May 1967 Continental Oil Company (CONOCO) became the first of the major oil companies to sign a production-sharing contract. The contract area was the Barito basin of South-East Kalimantan. The signature bonus which it had to pay was $1 million. Subsequently two other major companies—Union Oil and Sinclair Oil—signed, as partners, a similar agreement to cover onshore and offshore areas in North Sumatra and East Kalimantan. Altogether fifteen production-sharing contracts were signed in 1968, with some companies concluding more than one contract. The companies were:

Union Oil
International Oil Exploration
Phillips Petroleum Company of Indonesia
Total Indonesie
Indotex
Virginia International
IIAPCO
Agip
Continental Oil
Mobil Oil
Indonesia Frontier Petroleum
Javasea Oil
Indonesia Gulf Oil

Dr Ibnu Sutowo was able to obtain more than $18 million in signature bonuses from these contracts, as well as $54 million in production bonuses to be paid when production reached 50,000 to 300,000 barrels a day. In addition, most of the 1968 contracts had a clause which increased the government's share from 65 to 67½ per cent when production reached 65,000 or 70,000 barrels. In all Dr Ibnu Sutowo obtained a total commitment of $220.5 million to explore Indonesia's oil potential, mainly in offshore areas, from the 1967 and 1968 production-sharing contracts.

With the signing of the agreements the oil exploration companies in Indonesia operated under two forms of contractual arrangements with the government, viz. the older Contracts of Work of Caltex, Stanvac and Calasiatic Topco, and the new production-sharing agreements which all the other oil companies had to sign after the government had officially accepted the principle of production-sharing.

Acting President Suharto had, in the interim, reorganized his Cabinet, choosing the President of the University of Indonesia, Ir. Sumantri Brodjonegoro, to replace Ir. Bratanata as Minister of Mines. The Oil and Gas Affairs Department was put back under the control of the Ministry of Mines, a move which was designed to heal the rift between the Ministry and the Directorate-General of MIGAS.

The newly-appointed Minister was of the view that the two state oil companies, one mainly concerned with production and the other with domestic marketing, should be merged in order to maximize the use of the country's limited manpower, capital and resources. PERMINA and PERTAMIN were accordingly merged into a single national oil company, P.N. PERTAMINA (Perusahaan Negara Pertambangan Minyak dan Gas Bumi Nasional or National Oil and Gas Mining Company) through Law No. 28 of August 1968. A Board of Directors was formed with Dr Ibnu Sutowo as the President-Director.

The recent history of the Indonesian oil industry is very much a history of PERTAMINA and the policies it has pursued in developing the country's oil resources. Its ultimate aim is to conduct all Indonesian oil activities but up to now it had neither sufficient capital resources nor the trained local personnel to achieve this aim. As Dr Ibnu Sutowo has said, 'Although the Concession system had been terminated, we do not deny that we are still in need of foreign oil companies. We still need their skills, technology, capital and work experiences.' (*Indonesian Perspectives*, December 1974, p. 75.)

PERTAMINA itself has employed its limited resources to conduct exploration in relatively low risk areas, namely, onshore locations next to existing oil fields, while leaving the high risk exploration activities in offshore and isolated onshore areas to the private oil companies. Working on this basis PERTAMINA has made a number of oil discoveries, of which the most important is the Jatibarang field on the north coast of West Java. The field covers a large area from Cirebon to Banten in West Java. It came on-stream in 1973, and was soon producing an output of 40,000 barrels of low-sulphur crude a day.

In order to optimize its operations and maintain better contact with the foreign oil companies, PERTAMINA established five production units—Unit I in North Sumatra; Unit II in South Sumatra; Unit III in Java and Madura, Unit IV in Kalimantan, and Unit V in East Indonesia. Subsequently two other units were added—Unit VI in Central Sumatra; and Unit VII in Sambu Bland. Units I to V operate oil fields in their respective areas of responsibility. In addition PERTAMINA's research institute LEMIGAS produces some crude from its small fields in Java.

PERTAMINA received a considerable amount in payments from the foreign oil companies during the period between 1968 and 1970 when a large number of production-sharing contracts were signed. It used these funds to finance numerous projects, many of which were not oil-related, and some of which were not revenue-earning. Among these were a road-linking Plaju and Palembang in South Sumatra; improvements and extensions to Tanjung airfield in Kalimantan; tourist promotion projects; fully-equipped petrol stations; international class hotels and motels. It also donated equipment and funds to educational institutions in Indonesia and spent substantial sums on roads, schools, clinics and mosques for towns and villages in Java, Sumatra and Kalimantan. Apart from these, it established various subsidiary and joint-venture companies on a wide front—air transport, shipping, real estate, management services, data processing, oil trading, communications, insurance, petrochemical and steel industries, a fertilizer factory and even a padi estate. By taking on these additional non-oil commitments PERTAMINA eventually became the second biggest spender after the Indonesian government. But in so doing it assumed a role to which it could not devote its full attention, one markedly larger than simply being a national oil company concerned with the development of oil and gas resources.

In order to finance its multi-faceted activities, PERTAMINA has had to borrow heavily from international banking and other sources against future crude oil production, to the extent that alarmed banking circles supervising government development loans. Furthermore, many people within Indonesia, including highly placed government officials, had been concerned by the manner in which PERTAMINA, under Dr Ibnu Sutowo, had conducted its affairs, asserting that its growth was proceeding outside the framework of any governmental regulation, and that it was in fact 'a government within government'.

Up to then PERTAMINA had been subject to the provisions of a law passed in 1960 when conditions in the oil industry were quite different and the state oil company was a relatively small concern contributing little to the national budget. Evidently new legislation was needed to regulate the affairs of a rapidly expanding state oil enterprise now working within the context of an industry whose contribution to the government's budget had increased from only 5 per cent in 1966 to 33 per cent in 1971. Accordingly, President Suharto, after considering the report of a special Presidential Commission (Committee of Four), directed a team comprising the Ministers of Mines and Finance, the National Development Planning Council (BAPPENAS) and officials from other ministries to draft a new law to regulate the activities of PERTAMINA.

This law, known as the PERTAMINA Law (Law No. 8 of 1971), came into effect on 1 January 1972. It established a new corporation—PERTAMINA[1]—in place of P.N. PERTAMINA. It created a Government Council of Commissioners to determine PERTAMINA's general policy and approve its policies, budget, borrowings, establishment of joint ventures and subsidiaries and other transactions. The President retained his prerogative to appoint and discharge council members, as well as the final authority on all decisions. The management of PERTAMINA was vested in a Board of Directors, consisting of the President-Director (Dr Ibnu Sutowo) and five other directors, each the head of one of the five main divisions of the company—exploration and production; refining and petrochemicals; domestic marketing; shipping; and finance and administration.

The Law also laid down the conditions for the entry of PERTAMINA into fields and activities beyond the oil and gas sector. Thus the company could engage in such activities only with the President's consent and only if they were connected with oil and gas operations. But although the Law outlined the framework wherein the company could participate in non-oil activities, in practice a great deal of leeway was available for decisions to be made on the kinds of enterprises PERTAMINA entered into. The passing of the Law did not prevent PERTAMINA from receiving proposals ranging from shipbuilding and construction to tourist services. It was, in fact, PERTAMINA's over-commitment on its multiple ventures that was to make it the biggest corporate bank borrower in the developing world by 1974, and was to eventually precipitate it towards a major crisis affecting not only the company but also the Indonesian government.

[1] Perusahaan Pertambangan Minyak dan Gas Bumi Negara.

The policy of Dr Ibnu Sutowo had been to maximize oil revenues derived from the development of Indonesia's oil resources. One of the ways to achieve this was 'to ensure that the money used for the development of oil resources is mostly spent in Indonesia' (Ibnu Sutowo, 1972). This entailed the provision of the services and personnel needed for the exploration, production, processing and transportation of oil. PERTAMINA consequently built up its own tanker fleet, had an air service consisting of helicopters and fixed wing aircraft under the management of P.T. Pelita Air Service, operated refineries, and provided other services and facilities such as housing, land, insurance, communications, training centres as well as hospitals.

Other oil-related high-cost undertakings initiated by PERTAMINA were LNG plants, a new seaport in the Semangka Gulf in South Sumatra to serve mammoth tankers which have to bypass the Straits of Malacca, and a floating fertilizer plant that could be towed from gas field to gas field. The company, in a joint venture with Nissho Iwai Company of Japan and Pacific Bechter, Inc. of the U.S., also planned to develop Batam Island, 20 km south of Singapore, into a logistics and operational oil base for Indonesian oil operations, as well as a transhipment centre, and a new industrial estate.

PERTAMINA's position and reputation as an organization with managerial talent and resources served to attract to it many proposals to participate in ventures which lay outside the oil industry. One of these was a 20 000 ha padi estate in South Sumatra for which the company was to provide a capital of $150 million. Another even more ambitious and expensive project was the Krakatau steel plant in West Java, originally started with Soviet aid and suspended after Gestapu.[1] PERTAMINA entered into a joint 60/40 venture with the Ministry of Industry to complete the construction of the plant, the first stage of which was estimated to cost nearly $1 billion. In addition PERTAMINA entered into many lower-cost projects—schools, roads, a television station, mosques,

---

[1]The acronym for Gerakan tiga-puluh September or September 30th Movement. A group of military officers and political leaders calling themselves Gestapu attempted a coup against the government on 30 September 1965. The attempted Communist-inspired coup had widespread repercussions throughout the country.

government offices, utilities, bridges—from which it drew no financial returns and which, in normal circumstances, should have been undertaken by a development agency and not an oil company.

In the early years of its formation PERTAMINA financed some of the smaller projects from the substantial revenues it obtained from the oil companies who had signed the production-sharing contracts. But as both the number and the scale of its commitments increased, it has had to borrow heavily, mainly from international banks, to implement its projects. Within a short period of time PERTAMINA's loans had exceeded the ceiling on short- and medium-term commitments set by the International Monetary Fund (IMF) for Indonesia, so that the country lost its IMF stand-by drawing rights. It was only after Vice-President Agnew's visit to Indonesia in 1973 that U.S. aid to Indonesia was resumed (*Fortune*, 1973).

In order to avoid restrictions and governmental controls on medium-term loans of one to fifteen years PERTAMINA borrowed short-term loans from foreign banks, hoping to repay these from a $1.2 billion long-term loan it was negotiating for as well as from increased oil revenues. However the long-term loan fell through, while to compound the problem, oil production and revenues dropped because of reduced Japanese imports. Consequently the company could not repay its short-term loans totalling $100 million when these were due in February/March 1975, and a rescue operation had to be mounted, under Presidential instructions, by the Central Bank of Indonesia (Carlson, 1976). On 4 March 1976 the Indonesian government announced that Dr Ibnu Sutowo and four other directors of PERTAMINA would be relieved of their duties.

When the government conducted a survey of PERTAMINA's financial position it found that the company had incurred debts totalling about $7.6 billion, of which $1.4 billion was no longer the responsibility of PERTAMINA. Of the $6.2 billion debt directly incurred by the company, $3 billion was for the hire-purchase and charters on ocean-going and domestic tankers (official statement reprinted in *Sinar Harapan*, 21, 22 & 24 May 1976).

The events that led to the mismanagement of PERTAMINA and

the consequent financial crisis that have saddled Indonesia with debts that would by 1979, require 19 per cent of its export earnings to repay (according to the World Bank; quoted in *Asia Research Bulletin*, 30 November 1976) must be viewed in the light of Dr Ibnu Sutowo's conception of the company as one devoted not only to the development of Indonesia's oil resources but also as one concerned with the development of the Indonesian economy. Loans were raised from external sources in the expectation that the earnings from the non-oil enterprises coupled with those from oil would be sufficient to cover their eventual repayment. As it turned out Dr Ibnu Sutowo was incorrect on both counts, and instead of achieving his dual objectives of maximizing oil profits and accelerating the pace of Indonesian economic development, he only succeeded in imposing a massive financial burden on the country.

PERTAMINA after 1975

The 1975 PERTAMINA crisis was a major set-back not only to the national oil company but also posed serious problems for the country's balance of payments and its domestic development plans. The government instituted a series of measures to correct the situation, beginning with an official announcement by the Central Bank in March 1975 that the Indonesian government would meet payments on all of PERTAMINA's legitimate outstanding debts. The Bank of Indonesia provided $1.55 billion in foreign exchange to cover PERTAMINA's short-term loans. Special teams were appointed to renegotiate the company's domestic and overseas contracts. Payments of oil revenues by foreign oil companies had to be made direct to the Bank of Indonesia. Restrictions were placed on further PERTAMINA borrowing. The government also launched a series of moves to renegotiate or cancel previous tanker contracts.

The non-oil related activities of PERTAMINA were transferred to other government departments or agencies. The new strategy would be to divert the company of its numerous non-oil centres of earnings such as the Krakatau steel plant, the floating fertilizer factory, padi estate, high-rise buildings, etc., and concentrate solely on one earnings centre, namely oil. PERTAMINA's former role as a national development institution was accordingly abandoned.

PERTAMINA itself was re-structured in such a way that authority now flows vertically to six departmental directors who are responsible to the President-Director. The latter in turn reports to a government supervisory board which now wields real power instead of playing only a nominal titular role. The re-organization of the company was officially announced in Presidential Decree No. 44/1975 issued in December 1975. In March 1976, Major-General Piet Haryono, a military financial administrator, was appointed the new President-Director of PERTAMINA.[1]

Article 2 of Decree No. 44/1975 states that the main tasks of PERTAMINA are 'To execute the operation of oil and natural gas by gaining output as maximum as possible for the welfare of the People and State; to supply and save the demand on oil, natural gas, and fuel for the home country'. The new President-Director, in a press conference in October 1977 outlined the objectives of the company's operations:

1. To expand oil and gas exploration and production.
2. To increase oil exports and diversify export markets.
3. To save the domestic petroleum market, strengthening domestic tanker facilities and distribution networks.
4. To ensure that development of Indonesian oil is accompanied by the transfer of foreign technology to Indonesia.
5. To ensure that petroleum sector activity generates increased employment and upgrading of skills and training.
6. To see that the expansion of the petroleum sector contributes to the economic development of the outer islands (*Petroleum News*, December 1978).

The new role of PERTAMINA was described by President-Director Haryono in the following statement: ' . . . if there ever was a time in PERTAMINA history when PERTAMINA grew for its own sake without regard to the economic development of the Nation as a whole, that time is now past. PERTAMINA does not set development objectives—PERTAMINA responds to those objectives set by

[1] In April 1981 Drs Joedo Sumbono, Director of Domestic Supply, took over as President-Director of PERTAMINA when Major-General Piet Haryono retired at the end of his five-year contract.

the Government. PERTAMINA does not lead development—it supports it. PERTAMINA does not act in PERTAMINA's interest but in the interest of Indonesia' (quoted in *Pertamina Today*, 1979, p. 108).

Apart from these institutional rearrangements, the government also took action to revise its contracts with the foreign oil companies in a move to raise revenue to meet the debts incurred by PERTAMINA and to maximize its earnings from its oil resources. Earlier, in 1974, the new profit situations arising from the sharp rise in world oil prices had prompted it to modify its contractual arrangements with the oil companies on the grounds that the resultant windfall profits should accrue largely to the state and not to the oil companies. The Contract of Work agreements had provided for a 60/40 or 65/35 equity split in favour of PERTAMINA, cost recovery for the oil company, and a special payment of 'pro-rata' oil by the company to PERTAMINA for domestic consumption. A new profit-sharing formula was instituted in early 1974, replacing the earlier equity split. This earlier split was applied only to the first $5.00/barrel, and was tied to an inflation index. The revised conditions provided for the 60/40 or 65/35 split to be applied only to the 'base barrel' portion; the remainder would be split 85/15 in favour of PERTAMINA. A further modification was made in 1975 which provided that on any amounts realized in excess of the 'base barrel' the PERTAMINA share would increase to 85 per cent for production up to 150,000 barrels/day; 90 per cent for production between 150,000 and 250,000 barrels/day; and 95 per cent for production over 250,000 barrels/day. Companies with production-sharing contracts had their contracts similarly revised, with the split in the government's favour being increased to 85 per cent on the value of a barrel over $5.00. The 'base barrel' on which the 85/15 split came into operation was raised to $5.83 in June 1975. However, the 'two-tier' system was replaced by a straight split of 85/15 in favour of PERTAMINA in mid-1976, with various recovery allowances commensurate with reserves and production levels (*Petroleum News*, January 1980).

The renegotiation of its existing contracts with the oil companies to raise additional funds to meet the PERTAMINA crisis began in late 1975 and was completed in 1976. Talks were held initially with the companies operating under Contract of Work agreements (Caltex and Stanvac) for an additional surcharge of $1.00 for every barrel of oil produced in the contract areas. Both the companies eventually agreed to the terms and signed the revised contracts in early 1976.

In January 1976 the government also renegotiated its contracts with the foreign oil companies operating under production-sharing agreements. Changes were proposed for the cost recovery formula and profit split of the existing contracts so as to increase the governmental 'take'. The companies were slower to accede to the new terms and the deadline for acceptance had had to be extended from 31 July to 14 August 1976 before all of them signed the revised contracts.

Under the revised contracts the production-sharing companies would be divided into two groups—those with seven or less than seven years' oil reserves, and those with more. The first group would be allowed to recover their capital costs by depreciating over seven years, while the second group would be allowed to recover their capital costs over fourteen years. The double declining depreciating method was to be used in place of the original 40 per cent capital cost recovery 'off the top'. Provision would also be made for the recovery of non-capital costs by allowing the companies to expense these without limit in the year of expenditure, with those costs carried forward being recovered on a straight-line basis at 8 per cent interest. After deduction for costs, the oil proceeds or 'profit oil' would be split 85/15 in favour of the state. These new arrangements gave Indonesia, through PERTAMINA, an increase of $550 million in oil revenue.

Apart from the immediate need to service the debts incurred by PERTAMINA under its former President-Director, Indonesia has to increase its oil earnings because of its heavy dependence on them to finance its overall development efforts. The contract revisions discussed above were designed to maximize the governmental share of the production split of current and future production. Other than this, Indonesia has to look to increased production of its hydrocarbon resources and to upward revisions in oil price levels for higher oil revenues.

In the event production did record increases in 1976 and 1977 but the decline in productivity of the older oil fields which was not made up for by the discovery of significantly large fields resulted in a fall in production in 1978 and 1979. To a certain extent this fall in crude oil production was offset by the production and export of liquefied natural gas, which began in 1977.

However the main factor which has served to maintain governmental earnings in the oil sector despite the drop in oil output was the escalation in oil prices. As with all other oil producing and exporting countries in the world, the export earnings from oil have risen very substantially in Indonesia since October 1973 when the Arab oil embargo started and oil prices quadrupled. The background to this goes back to 1962 when Indonesia joined the Organization of Petroleum Exporting Countries (OPEC). OPEC was formed in September 1960 by Iran, Iraq, Kuwait, Saudi Arabia and Venezuela to combat the major oil companies' efforts to lower the posted prices on which their oil agreements were based. At that time the five countries were supplying about 80 per cent of the oil in the international market. In 1967 OPEC adopted its 'Declaratory Statement of Oil Policy' in which OPEC members would pursue an agreed-upon oil policy, with the ultimate aim of placing every oil enterprise in each OPEC country under that country's control.

Although Indonesia had raised its oil prices several times in the 1960s after negotiations with the oil companies, prices remained below the $2 a barrel level until the second quarter of 1971 when the average price of Indonesian crude went up to $2.17 a barrel. By the last quarter of 1973 the average price had increased to $6.16 a barrel, compared with $2.82 in the first quarter of that year.

When the Middle Eastern OPEC countries raised their posted oil prices from $5.16 to $11.60 a barrel in December 1973 Indonesia also raised the export price of its crude oil to $10.80 a barrel. Thus in the space of one year the price of Indonesian oil had increased by 380 per cent.

The rationale behind the increase was explained by Dr Ibnu Sutowo in two addresses he gave in 1974. The main points he made were: first, the availability of abundant cheap imported oil was one of the main reasons for the 'tremendous boom' of the industrialized countries after the Second World War. In effect the oil producing countries of the developing world were subsidizing the continued industrialization and growth of advanced countries by accepting unrealistically low prices for their oil.

Second, the pricing of a non-renewable natural resource such as oil should be value-based rather than cost-based. Cost-based pricing did not provide a fair trade value for natural resources. Third, up to 1970 the major international oil companies had the power to determine production, development, exports and prices in oil exporting countries while also receiving a larger share of the profits. But OPEC had succeeded in transferring the powers of economic decision-making from the companies to the governments of the producing countries. Consequently the OPEC countries could use their economic leverage to adjust oil prices to a level which more adequately reflected the value of their resources and which compensated them for the inflationary increase in the prices of goods and services which they must import from the industrialized nations.

Fourth, Indonesia could not accept a price for its irreplaceable natural resources which was below their value as to do so would be detrimental to its development efforts. It badly needed the funds which the increased oil prices would bring in in order to raise the Indonesian standard of living above subsistence level (Ibnu Sutowo, 1974a, 1974b).

In 1974 Indonesia raised the export prices of its oil twice—from $10.80 to $11.70 a barrel at the end of March 1974; and from $11.70 to $12.60 a barrel in July 1974. These increases were not in line with those of the other OPEC countries as Indonesia had based its price calculations on the fact that its oil was of a better quality than Middle Eastern oil, its marketing region and transportation costs were also different from those in the Middle East, and its oil price was in the form of 'realized' price and not 'posted price'.

In October 1975 Indonesia adopted a four-level pricing system for its various grades of oil. The first grade was Rantau crude, at $13.00 a barrel, followed by Minas, Attaka, Sepinggan, Duri, Bekapai, Badak, Bunyu and Lirik crudes at $12.80 a barrel, Cinta, Jatibarang and Handil crudes at $12.40 a barrel, and lastly Walio and Pamugian Juata/Tarakan crudes at $12.10 a barrel.

In December 1976 at the Qatar conference OPEC agreed, by a majority, to raise oil prices by 10 per cent as from 1 January 1977 and a further 5 per cent in July 1977. Saudi Arabia, which produces about 30 per cent of OPEC's total oil production, and the United Arab Emirates decided to increase their oil price by only 5 per cent. Indonesia went along with the majority decision, and accordingly revised its prices upwards, though the weighted average of its price increases was about 7 per cent instead of 10 per cent. The reason for this difference was the adoption of a new pricing system centred on Indonesia's bench-mark Minas crude. At $12.80 a barrel in 1976 the Minas crude was regarded as being over-priced. To correct this and to make it more competitive with crudes produced elsewhere the price of Minas crude was lowered by 44 cents a barrel. To the new price of $12.36 a barrel was added the 10 per cent price increase based on Saudi Arabian bench-mark crude, amounting to $1.19. The final result was that on 1 January 1977 the price of Minas crude went up from $12.80 to $13.55 instead of $14.08, corresponding to an actual upward revision of 5.8 per cent. In contrast, those crudes that were regarded as being under-valued had their prices adjusted by more than 10 per cent. Thus the price of Attaka crude was raised from $12.80 to $14.10 a barrel, an increase of 10.1 per cent; while that of Attaka (a North Sumatran light crude) went up by 10.7 per cent, from $13.00 to $14.40 a barrel.

A period of relative price stability in 1977 and 1978 was followed by a flurry of price increases in 1979 and 1980 as a result of the new OPEC pricing policies and the shortfalls in Iranian output which affected the world supply situation. The end result up to May 1980 was an average price increase of Indonesian crude oil of 139 per cent over the average price in January 1978 (Table 2.1).[1] Whatever the short-term fluctuations in oil prices, oil analysts appear confident that oil demand will be constrained by tight supplies in the 1980s (*Petroleum Economist*, February 1980).

The final outcome was that in spite of production declines the foreign exchange earnings of Indonesia were boosted considerably, entirely because of the large price increases of 1979 and 1980. Gross foreign exchange earnings from oil were about $7 billion in 1978; in

[1]In December 1980 the OPEC countries agreed at the Bali meeting that members could charge prices ranging from $32 to $41 a barrel for their oil.

TABLE 2.1
OFFICIAL PRICES OF CRUDE OIL,
INDONESIA, 1978–1980
(in US$ a barrel, f.o.b.)

Name	January 1978	January 1979	January 1980	May 1980	Percentage over January 1978
Arjuna	13.70	14.40	28.95	32.95	140.5
Attaka	14.10	14.95	30.25	34.25	143.0
Badak	14.10	14.95	30.25	34.25	143.0
Bekapai	14.10	14.95	30.25	34.25	143.0
Cinta	13.15	13.50	27.00	31.00	136.0
Jatibarang	12.80	13.15	26.60	30.60	139.0
Handil	13.30	13.95	27.55	31.55	137.0
Minas	13.55	13.90	27.50	31.50	132.0

Source: Pertamina Bulletin, various issues.

1979 they had increased by over 40 per cent, to about $10 billion. This was 70 per cent of the total gross foreign exchange earnings of Indonesia, and was the largest single source of such earnings in the country (American Embassy, 1980, pp. 9–10).

This has produced a balance of payments surplus situation, and has helped the government to overcome the financial burden of PERTAMINA's debts. PERTAMINA's own indebtedness to the government and the Bank of Indonesia was gradually reduced through contractual settlements, through payments for its projects assumed by other governmental agencies and through the acquisition of some of its assets by the government. By 1979 PERTAMINA's short-term debts and tanker debts had virtually been settled. By February 1981 the national oil company's outstanding debts had been reduced to $2.4 billion (Dr Subroto, Minister of Mines and Energy, quoted in the *Straits Times*, 11 February 1981).

PERTAMINA and Indonesia have successfully surmounted the 1975 financial crisis, and the oil company is now in a position to carry out its original mandate of assuming responsibility for all petroleum activities in Indonesia. This includes exploration, production, refining, transportation and marketing. In so far as production is concerned, PERTAMINA has up to now been responsible for only about 5 per cent of the total Indonesian output, the rest being from the 37 or so foreign oil contractors which the national oil company oversees. But the indicators are that PERTAMINA's involvement in exploration and production will increase in the 1980s as a result of the $310 million oil exploration loan from the Japanese National Oil Corporation (JNOC) as well as the separate JNOC loans for the development of existing oil and gas fields in PERTAMINA areas.

III

Geological Setting for Indonesia's Petroleum Deposits

PETROLEUM or 'rock oil' is a naturally occurring mixture of hydrocarbons—liquid, gaseous, and solid—which forms the basis of what is known commercially as crude oil, natural gas and natural asphalt. It is widely accepted that the formation of oil is a very slow process, and commercial oil is rarely found in rocks less than 1 million years old.[1] Petroleum occurs in rocks of all ages, from Pre-Cambrian to Pleistocene, a time span of over 500 million years. It also occurs in all the major sedimentary rock types, with shales and clays being the most common petroleum source rocks.

Oil fields are universally associated with sedimentary areas. It is generally believed that petroleum originates from organic matter entombed in sediments in the course of their formation. The generation of hydrocarbon from such organic matter is postulated as a cooking process, whereby the kerogen-containing sediments are 'cooked' during deep burial. The rate of subsidence will determine both the amount and duration of heat generated for the cooking process. As such cooking proceeds, carbon dioxide, methane, ethane and progressively larger hydrocarbon molecules are generated in gradually increasing quantities (Kingston, 1978).

A minimum depth of burial appears to be necessary for petroleum to form. One authority (see Hobson, 1954) cites a minimum thickness of 1 500 m (5,000 ft) of sediments for petroleum to form and commence to migrate into reservoirs. Since the deposition of

[1]In the 1950s sediments cored from the Gulf of Mexico, and established by radio-carbon dating to be 10,000 years old were discovered to contain over 1 per cent of petroleum hydrocarbon. This means that petroleum compounds are formed soon after burial.

sediments by natural processes is usually extremely slow, the estimated time needed for the formation of 1 500 m of such sediments as shales and sandstones is in the order of 2 to 4 million years. However, as should be expected, the greater the rate of sedimentation the quicker will oil be formed. Such indeed is the case, and there is a close correlation between large hydrocarbon accumulations and areas of thick and rapid sedimentation, usually deposited under marine conditions.

The conditions which favour rapid sedimentation are best found in the orogenic belts of the earth, more specifically, in areas where there is a fast subsiding sea floor beside a land mass being forced upwards by earth movements during the process of mountain building. In such areas, known in the terminology of classical tectonic theory as geosynclinal basins, sedimentation rates can be as high as 600 m (2,000 ft) per million years, and the thickness of sediments accumulated in a single geologic period can be 9 000 m (30,000 ft) or more (Stebinger, 1950).

In Indonesia the most important oil-bearing and oil-producing basins belong to the Tertiary period, although oil and gas seepages have been found in Permian and Mesozoic rocks in the islands of Timor and Seram in eastern Indonesia (Audley-Charles & Carter, 1974). However, the oil prospects of the pre-Tertiary are rated as poor because these pre-Tertiary rocks mostly comprise intensely folded or metamorphosed sediments or are igneous rocks.

Geological conditions in Indonesia are complex, with stable continental blocks, volcanic and non-volcanic island arcs, and sedimentary basins of varying dimensions. It is also a region of strong crustal disturbances as evidenced by recurrent earthquakes, earth tremors and volcanic eruptions, which have occurred in one part or another of the volcanic island arcs throughout the entire Tertiary period to the present-day.

Until recently geologists have attempted to explain the geological evolution of Indonesia in terms of the classical tectonic theory, whereby an initial period of geosynclinal subsidence and sedimentation would be followed by a phase of intermittent orogeny (mountain building) and magmatism, characterized by volcanism and the emplacement of granitic rocks (see, for example, Westerweld, 1952).

However, the new plate tectonic theory visualizes the earth's crust as consisting of a number of large and small rigid plates moving relative to other adjacent plates—either converging, diverging or sliding past one another. The margins of the plates are usually zones of instability, characterized by the deformation of crustal rocks by thrusting, folding and faulting, and by seismic and volcanic activity. This plate tectonic theory is now commonly applied to explain the geology of Indonesia (see Hamilton, 1973; Katili, 1973; Gribi, 1973).

According to Hamilton (1973) the Indonesian-Melanesian region is the boundary zone of three large lithospheric plates—the Asian, the Australian-Indian Ocean and the Pacific plates—converging obliquely. Throughout most of the Cenozoic era the Australian-Indian Ocean plate has been moving northward relative to the Asian plate, while the Pacific plate's motion has been W-NW. The relative motion of the Asian continental plate where it collides against the Australian-Indian Ocean plate is horizontal, while that of the oceanic plate is downward at an inclined angle, as it under-rides (subducts) the continental plate. The boundaries of the subduction zone are marked by the submarine trenches (Sumatra-Java trench, Timor trench, etc.) of the Sumatra-Java-Timor-Outer Banda-Seram island arcs.

Katili (1973) has constructed a plate tectonic model for western Indonesia, using as his base the double island arc model of J. H. F. Umbgrove. The zones inward across the continental plate are: the submarine trench, marking the subduction zone; the non-volcanic outer island arc of the Mentawai islands; the inter-deep; the magmatic or volcanic inner island arc of Sumatra, Java and the volcanic islands east of Java; and the foreland basin.

The model for eastern Indonesia is more complicated. Gribi (1973) considers the Moluccas as the meeting zone of four crustal plates: the Pacific plate on the north-east moving south-west; the Indian Ocean plate on the south-west pushing north-east; the Australian shield providing a relatively stable eastern buttress; and the Sunda Shelf marking the western limit of the area. The collision of these plates has resulted in a gigantic anti-clockwise sworl (see also Katili, 1975; Hamilton, 1979).

In so far as petroleum is concerned, the foreland basins appear to

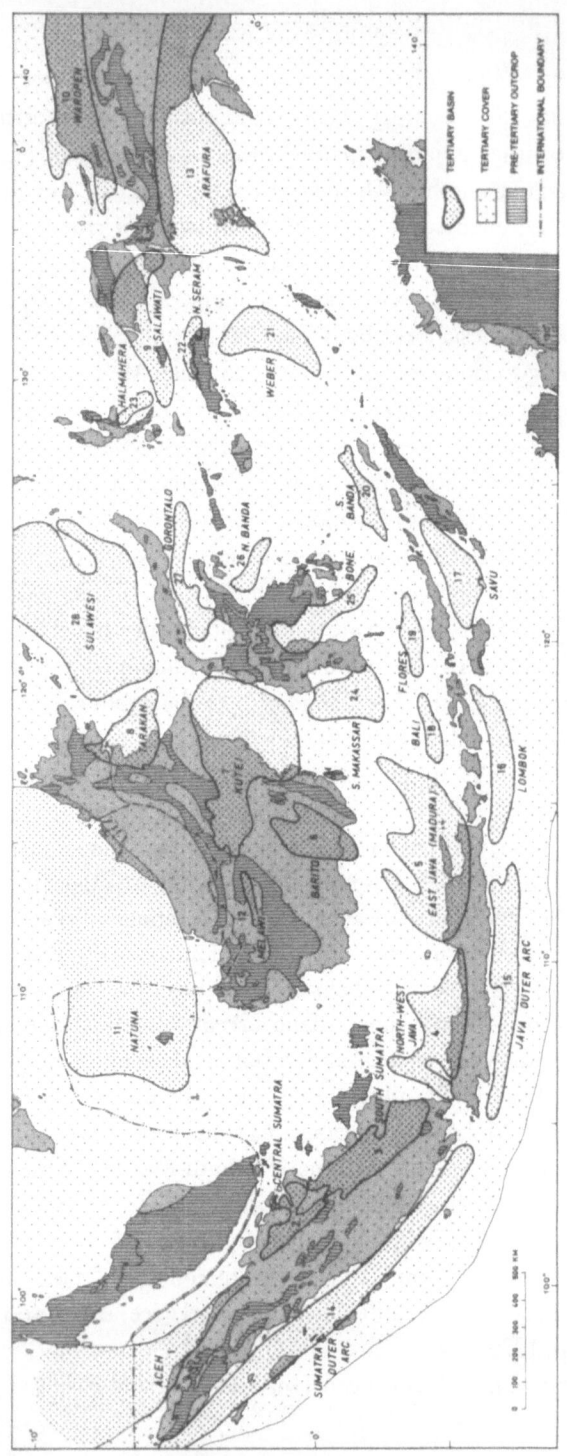

Fig. 3.1 The Tertiary basins of Indonesia
Source: Based on Fletcher and Soeparjadi, 1976.

hold the best prospects. They occupy the belt between the volcanic inner arc and the stable continental blocks of the Sunda Shelf and the Sahul Shelf, and were the zones of rapid sedimentation in Tertiary times when beds up to 12 000 m thick were laid down. The Tertiary foreland basins of western Indonesia and northern Java have in fact been proved to contain significant quantities of petroleum (Katili, 1973). Of lesser interest to the petroleum prospector are the smaller sedimentary basins in the belt between the non-volcanic outer arc and the volcanic inner arc (the inter-deep). Oil and gas seeps are present in these basins, although so far no significant accumulations have been discovered.

Figure 3.1 shows that a high percentage of Indonesia and its surrounding seas contain Tertiary sedimentary basins. The Tertiary is nearly everywhere unconformably overlying a Mesozoic eroded and weathered surface, except in eastern Indonesia where a continuous upper Mesozoic to Eocene sequence is present. Oil shows have occasionally been recorded from Mesozoic rocks, and, as well, the extensive asphalt deposits of south-eastern Sulawesi are found in Mesozoic sandstones.

Figure 3.1 also shows the main Tertiary sedimentary basins of Indonesia as classified by Fletcher and Soeparjadi (1976). These basins appear to have gone through a megacycle as well as two or more sub-cycles of sedimentation. Briefly, the megacycle consisted of a fully marine sequence during which the petroleum source rocks were laid down, sandwiched by transgressive and regressive phases during the Tertiary was followed by moderate to strong folding at the from conglomerates to sandstones, were deposited. Sedimentation during the Tertiary was follwed by moderate to strong folding at the end of the Tertiary, resulting in a gently folded thick prism of sediments, holding good petroleum prospects. All the important oil-producing basins of western Indonesia appeared to have gone through this cycle of evolution and sedimentation, although individual basins may show variations in the timing of the events. The megacycle ended at the end of the Pliocene in some basins but in others it continued into the Pleistocene and even into Recent time (Koesoemadinata & Nelson, 1970).

Oil migrated from the source rocks to accumulate in either the

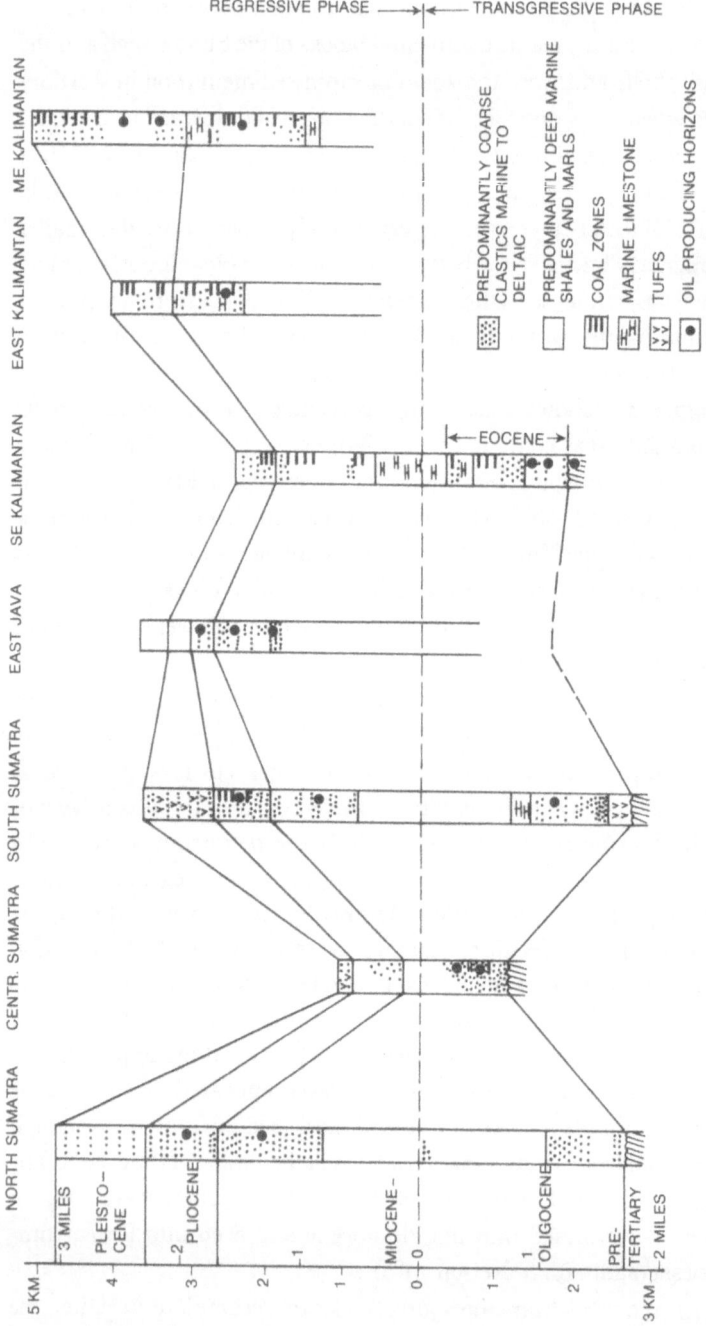

Fig. 3.2 Tertiary stratigraphy of the West Indonesian oil basins

transgressive or the regressive sands, that is, in either the lower part or the upper part of the Tertiary stratigraphic column (Fig. 3.2). Very occasionally, as in Kluang (South Sumatra) and some fields in East Kalimantan, both series may yield hydrocarbons. When exploration and production began in Indonesia at the end of the nineteenth century attention was focused on the sand reservoirs in the upper regressive series which were at depths accessible to the drill. Oil fields were developed in the upper regressive series in North-East and South Sumatra, East Java, and Kalimantan (Fig. 3.2). In these areas the lower lying formations of the transgressive series were either too deep for the drill then or were absent in that part of the sedimentary basin.

Data from West Indonesia indicate that stratigraphic conditions favour the trapping of oil in the upper regressive series in the deeper part of the basins, that is, the parts close to the volcanic inner arc. In the North and South Sumatra basins, for example, oil is produced in commercial quantities in the regressive facies in the deeper basinal parts. In the shallower parts overlying the stable Sunda Shelf the

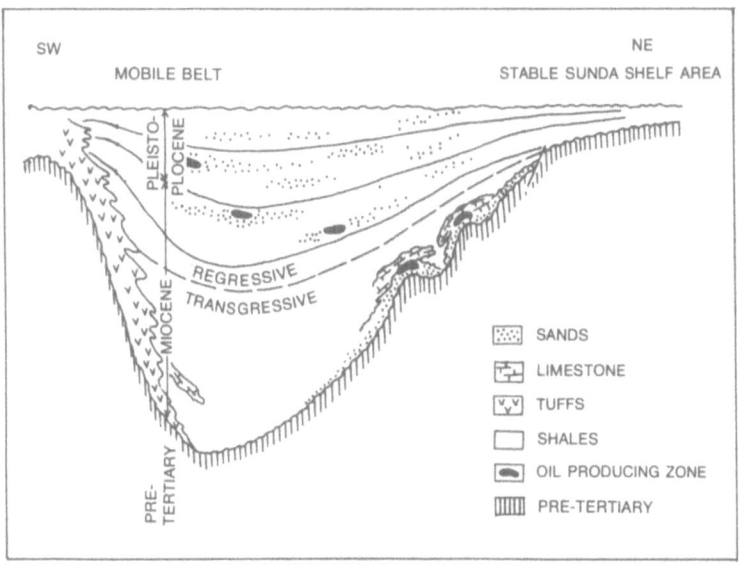

Fig. 3.3 Generalized stratigraphic diagram of the South Sumatra basin
Source: After Koesoemadinata and Nelson, 1970.

regressive facies is barren whereas the transgressive facies is productive, as seen in the South Sumatra basin (Fig. 3.3). Conditions in this case appeared to have favoured the migration of oil laterally from the source rocks into the lower Tertiary reservoir rocks of the transgressive phase (Fig. 3.2 & 3.3).

The discovery of oil in the lower transgressive facies in South Sumatra in 1918 led to an extensive search for oil in these formations throughout western Indonesia. These exploration efforts were intensified after the Second World War, and have resulted in the discovery of oil reserves larger than those contained in the upper regressive facies.

Soeparjadi *et al.* (1975) recognize a single major cycle of sedimentation in Indonesia during the Tertiary, with specific areas exhibiting marked variations in response to differential movement and collisions of bordering cratonic and oceanic plates. The cycle started in the early Tertiary (Eocene time) with a major marine transgression which laid down sediments over an irregular basement complex filled earlier in places by mainly non-marine sediments. This phase continued through the entire Oligocene and came to an end in the Middle Miocene when many of the oil-producing basins of western Indonesia were subjected to epeirogenic uplift. Major transform faults, tilting and large-throw adjustment rifts were associated with this event.

Following this general uplift, a regressive phase set in which continued until the late Pliocene when the collision of the Eurasian and Indian-Australian plates resulted in another orogeny. This Plio-Pleistocene tectonic period affected not only western Indonesia but also greatly modified areas in central and eastern Indonesia. Subsequent to this a second depositional cycle was initiated in the Quaternary when the seas again transgressed the basins.

More recently Beddoes (1980) has identified a common pattern of cycles of Tertiary sedimentation throughout South-East Asia (including most of Indonesia). These cycles are:

1. A Paleogene (Eocene to Early Oligocene) transgressive cycle. During this period the seas transgressed the basins peripheral to the Sunda Shield, while the interior basins received non-marine sediments. In Indonesia the only oil accumulation found in strata of

this age is the Tanjung field of the Barito basin, but in mid-1977 oil was tested from an Eocene carbonate reservoir in the East Java Sea, holding promise of further discoveries in Paleogene reservoirs in other basins. This cycle closed with the uplift of the core of the Sunda Shield at the end of Early Oligocene time, when the peripheral basins were filled up with regressive sediments. This brief regressive period was followed by:

2. A major Neogene (Late Oligocene to Middle Miocene) transgressive cycle. During this period many of the earlier regressive sediments were reworked. Together with the sandstones deposited they form the most prolific oil-productive reservoirs in South-East Asia, accounting for 54 per cent of the cumulative oil production up to mid-1979. Of lesser importance are the transgressive Neogene carbonates (2 per cent of cumulative production). In Indonesia the giant Minas field is the prime example of the productivity of the clastic reservoirs of this age, while the Arun gas/condensate field in North Sumatra is an example of a very large hydrocarbon accumulation in an Early to Middle Miocene carbonate complex. This transgressive cycle gave way to:

3. A Neogene (Middle Miocene to Early Pliocene) regression cycle caused by a major tectonic event. Local variations in tectonic setting resulted in differences in which this event was expressed in the peripheral basins of the Sunda Shield. In Sumatra, for example, the event was associated with intrusion and vulcanism. This cycle was interrupted by a Late Miocene transgression of brief duration when marine shales were deposited over the regressive sand reservoirs in some basins. The period (Late Miocene to Early Pliocene) which followed was one of maximum regression in South-East Asia. This was onlapped by an Early Pliocene transgression, after which a period of tectonic activity (uplift) brought in another regressive cycle which lasted until Early Pleistocene time. Petroleum production in South-East Asia from clastics from the Neogene regression amounted to 39 per cent of the cumulative production to mid-1979 of 10.68 billion barrels, and from carbonates 2 per cent. In Indonesia oil has been found in sandstone reservoirs of this cycle in the North Sumatra, South Sumatra, West Java Sea, Kutei and Tarakan basins, while the Neogene reef carbonates in Irian Jaya have also produced oil.

The Tertiary Basins of Indonesia

As the oil prospects for Indonesia lie mainly in the Tertiary, this section will be devoted to an examination of the Tertiary basins of the country. Soeparjadi *et al.* (1975) using current plate tectonic theory, classified the Tertiary basins of Indonesia into four types: (1) outer arc basins or arc front basins which lie between the non-volcanic outer arc and the inner volcanic arc of the Indonesian island arcs; (2) foreland basins or back arc basins, which lie between the volcanic inner arc and the craton; (3) interior cratonic basins which lie entirely on continental crust and have undergone a single major cycle of Tertiary deposition; and (4) open shelf basins on a continental margin such as the Bintuni and Salawati basins of Irian Jaya, the East and West Natuna basins, and the Kutei and Tarakan basins of Kalimantan.

Fletcher and Soeparjadi (1976) identified twenty-eight separate Tertiary basins in Indonesia and classified these according to their genetic characteristics into outer arc, foreland, cratonic and inner arc basins. Although they merge with each other to form the Tertiary cover which blankets most of Indonesia, the economically important basins are easily recognizable and belong to one or the other of the four basin categories. These basins were formed at continental margins, on foundered cratons, and along volcanic arcs adjacent to subduction zones as a result of structural changes associated with plate movements in pre-Tertiary times.

Nayoan *et al.* (1979) and Hariadi (1980) have more recently classified the Tertiary basins along the lines followed by Soeparjadi *et al.* (1975), with the addition of a new category of 'unspecified' basins for those basins where not enough is known of the geology to fit them into the other categories. The boundaries of these basins are defined by pre-Tertiary outcrops, Tertiary sediment thickness of about 1 000 m, or by volcanic cover. The distribution of the forty basins so recognized is shown in Figure 3.4. The marked differences of basin outlines between Nayoan *et al.* (1979) and Fletcher and Soeparjadi (1976) are due to the availability of geological control and definition criteria for the basins.

A long period of petroleum exploration and knowledge acquired in

Fig. 3.4. Tertiary basin types in Indonesia
(basin names are listed in Appendix B)
Source: Nayoan *et al.*, 1979.

tapping the hydrocarbon potential have brought to light a number of important geologic attributes of the Tertiary basins of Indonesia. First, there appears to be a correlation between the age of the Tertiary sediments and their hydrocarbon potential. Ninety per cent of all the oil and gas discovered have come from rocks of Middle and Lower Miocene age.[1] Such rocks make up about 20 per cent of the total volume of Tertiary sediments in Indonesia, and are regarded as having the greatest hydrocarbon potential, since they contain abundant source beds and related reservoir rocks. In contrast, the younger basins with a post-Upper Miocene history have no significantly large oil accumulations and appear to have only moderate hydrocarbon potential, although they contain the bulk of the Tertiary sediments.

Second, there is no correlation between basin size and abundance of hydrocarbon accumulations, as the small basins in Indonesia have proved to be as prolific as the large ones.

Third, there also appears to be no correlation between the thickness of the sedimentary columns and hydrocarbon accumulation. Central Sumatra, for example, has a very thin sedimentary column but is unusually rich in hydrocarbons. North-West Java, too, has a thin sedimentary column, but abundant proved hydrocarbon accumulations.

The twenty-eight Tertiary basins identified by Fletcher and Soeparjadi (1976) include onshore basins, onshore basins with offshore extensions, and offshore basins. Ten of the onshore basins are known to extend offshore: the North Sumatra (Aceh) basin (No. 1 in Fig. 3.1); the North-West Java basin (No. 5); the East Java basin (No. 10); the Barito basin (No. 12); the Kutei basin (No. 14); the North-East Kalimantan basin (No. 15); the North Seram basin (No. 28); the Salawati—Bintuni basin of the Vogelkop (No. 24); the Arafuru basin (No. 23); and the Waropen basin of Irian Jaya (No. 25).

Until 1966, exploration activities and the production of oil in Indonesia were from onshore fields (except for the offshore Bula field in Seram, which was exploited before the war). The increased interest

[1] In South-East Asia as a whole 56 per cent of the cumulative oil production up to mid-1979 was from rocks of Late Oligocene to Middle Miocene age, laid down during the major Neogene transgression (Beddoes, 1980, Table 19).

in oil exploration shown by oil companies since 1966 has resulted in the extension of the search for hydrocarbons to offshore areas throughout Indonesia, and a steady accumulation of data on the geology of offshore basins. Much of such data, however, are still unpublished, and lie in the confidential files of the companies and the archives of PERTAMINA.

The Indonesian offshore areas can be divided into the continental shelf, the continental slope and rise, and the abyssal plains. Estimates of the areal extent of these physiographic provinces are given in Table 3.1 below. The distribution of these areas is shown in Figure 3.5.

It will be seen that the greatest areal extent of continental shelf is in two geographically distinct regions: (1) the region between Peninsular Malaysia, Sumatra, Java and Kalimantan, and (2) the region between Irian Jaya and Australia. Since about three-quarters of the Indonesian continental shelf is covered by Tertiary deposits, some of great thickness, the hydrocarbon potential of this physiographic province is considered by geologists to be very attractive.

The discovery of major offshore fields in other parts of the world in locations where the onshore areas were barren has disproved the theory that offshore oil could only be found opposite onshore fields. This has stimulated offshore exploration in parts of the Indonesian continental shelf other than those opposite onshore fields. A

TABLE 3.1
OFFSHORE AREAS IN INDONESIA

Physiographic Province	Approximate Depth Range (m)	Estimated Area (sq. km)	Estimated Area with Tertiary Deposits (sq. km)
Continental shelf	0–200	1 900 000	1 400 000 (74%)
Continental slope & rise	200–5 000	2 700 000	2 000 000 (74%)
Abyssal plains	below 5 000	400 000	—
Total		5 000 000	3 400 000 (68%)

Source: Akil and Nayoan, 1973.

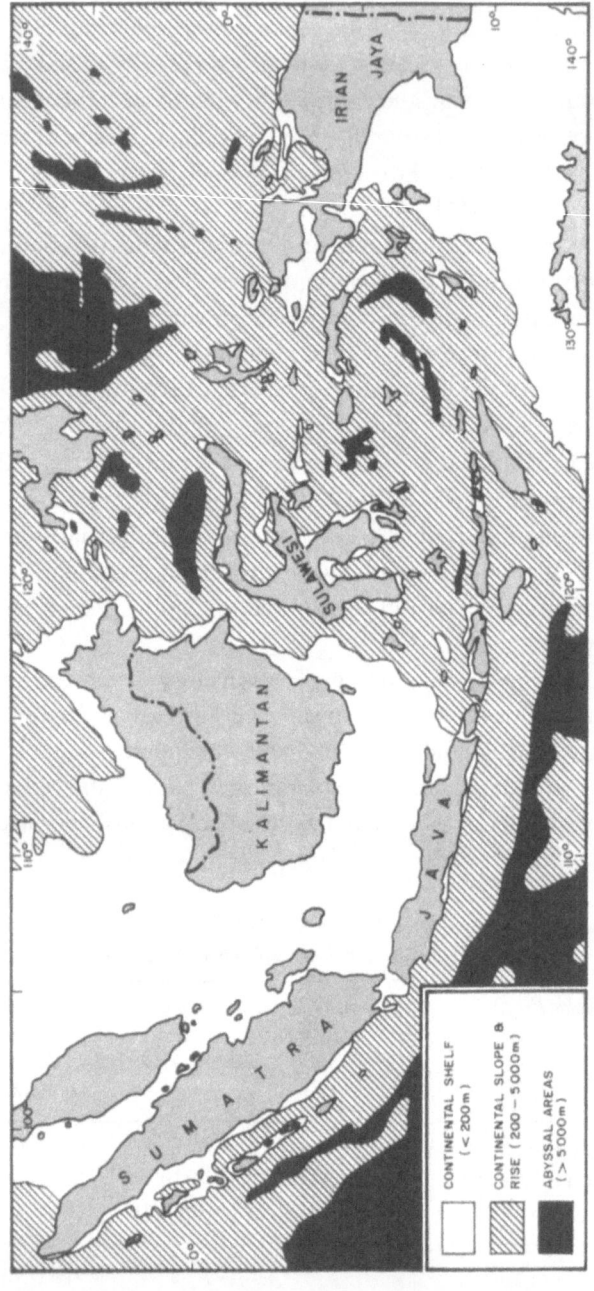

Fig. 3.5   The continental shelf, slope and rise, and abyssal areas of Indonesia
Source: Akil and Nayoan, 1973.

consequence of this was the discovery, through marine seismic investigation and subsequent drilling, that the Sunda Shelf was not a vast plate of granitic basement, of which the islands of Natuna, Anambas, Banka and Billiton were emergent parts, but rather a series of submerged Tertiary basins separated by islands of granite and older Mesozoic rocks, these islands being the emergent part of geanticlines on up-faulted rocks (King, 1971). An aeromagnetic profile across the South China Sea from Singapore to Sarawak indicates the possibility of a shallow narrow basin. The islands of Bangka and Billiton probably mark the boundary between the northern basin and the petroliferous Java Sea basin, which in turn is separated from the onshore South Sumatra basin. It is also possible, though not yet proved, that the delta of the Kapuas River extends from the western Kalimantan coast into an offshore basin between the Anambas and Natuna islands.

The continental slope and rise of Indonesia cover an even more extensive area (2.7 million sq. km) than the continental shelf. About three-quarters of the slope and rise is blanketed by Tertiary sediments of largely unknown thickness, although research elsewhere has shown that the margins of large continents usually have thick sedimentary successions. Turbidite sands, which can be good reservoir rocks, are known to occur on the Indonesian continental slope, but no other evidence is as yet available of the presence of hydrocarbon accumulations, and the petroleum prospects for the Indonesian continental slope and rise will become clearer only through further geological and geophysical investigation and wildcat drilling.

The major Tertiary onshore and offshore basins of Indonesia with their realized and unrealized hydrocarbon potential (Figs. 3.6 & 3.7) are described below under the four genetically-based categories of Fletcher and Seoparjadi (1976).

Foreland Basins

Foreland basins are formed along the edges of cratonic plates when these plates collapse or tilt under regional cratonic stresses or sedimentation pressures. The size and shape of the basins would vary according to the characteristics of the craton, the distance of the

58

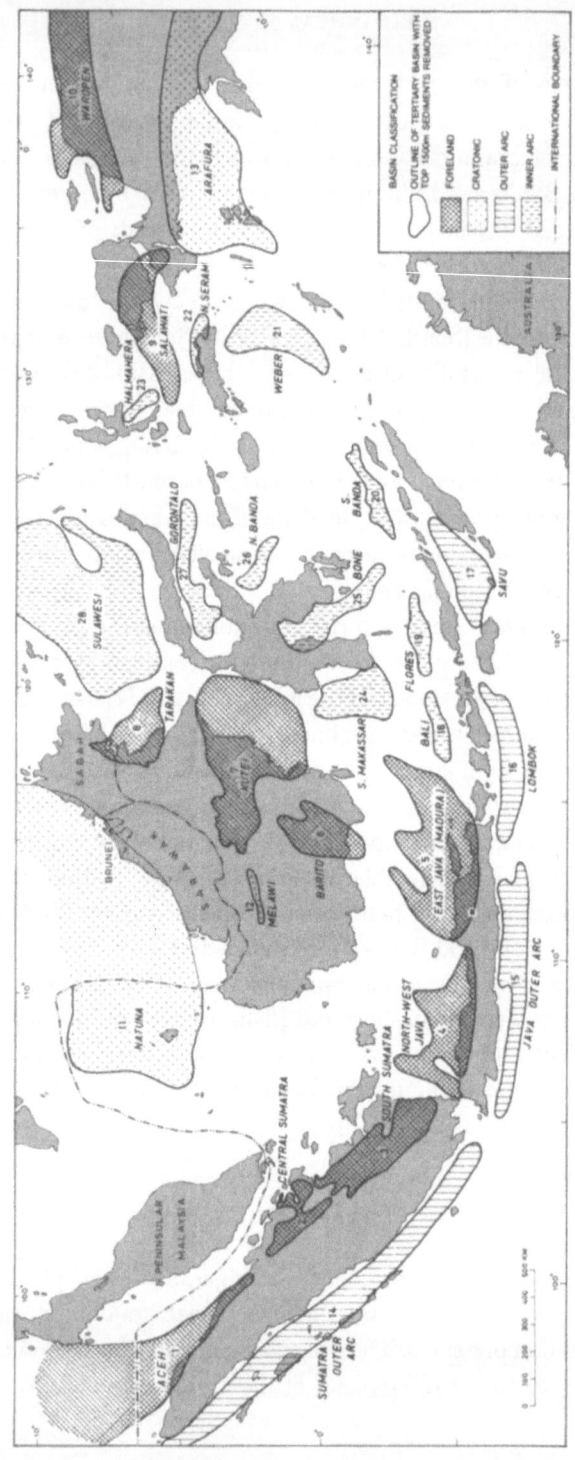

Fig. 3.6 Tertiary basin types in Indonesia
Source: Fletcher and Soeparjadi, 1976.

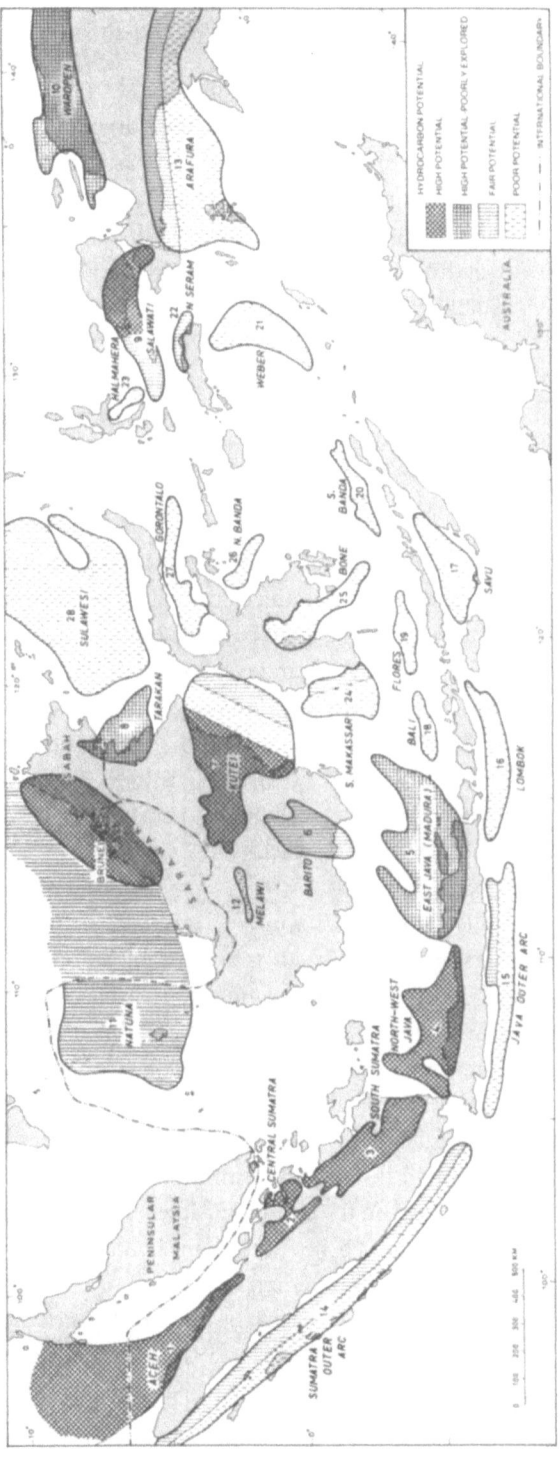

Fig. 3.7 The hydrocarbon potential of the Indonesian Tertiary basins
Source: Based on Fletcher and Soeparjadi, 1976 and PERTAMINA.

downwarp from the continental boundary and the amount of sediments deposited in the basins. In the foreland basins of Indonesia, as in the case of many other foreland basins in other parts of the world, such sediments are largely deltaic to neritic sands, silts and shales containing abundant organic material. These are prolific source rocks for hydrocarbons, so that the hydrocarbon potential of such basins is usually high to very high.

The foreland basins of Indonesia border the eastern edges of the Malaysian craton and the northern edges of the Australian craton. Apart from sediments derived from the cratons, the basins also received thick deposits of Late Oligocene and Lower Miocene carbonates laid down in shallow shelf areas. The hydrocarbon potential of these basins is considered to be very good, with abundant source rocks interbedded with sandstone reservoir rocks. In addition, the carbonate deposits containing reefs and calcarenite banks are regarded as having high promise for large hydrocarbon accumulations.

Most of Indonesia's production and reserves are from the foreland basins in Sumatra, North-West Java, the east coast of Kalimantan and Irian Jaya. These basins, especially those in Kalimantan and Irian Jaya, have not been fully explored. It is likely that with the employment of new seismic techniques substantial additional potential will be discovered in the basins.

ACEH BASIN (NO. 1 IN FIG. 3.6)

The geologic history of this basin is similar to that of the other foreland basins of Central and South Sumatra. Low basement arches separate the basins from each other. The three basins together form part of the foreland downwarp around the Malaysian craton. The Aceh basin has an area of about 60 000 sq. km. It is roughly triangular in shape, and as in the case of the other Sumatran basins, is bounded on the south-west by the Barisan Mountains, on the north by the Andaman Sea, and on the east by the foreland of the Straits of Malacca. The Asahan arch is believed to be the boundary separating the Aceh basin from the Central Sumatra basin.

Sedimentation in the Aceh basin began in the Eocene and continued throughout the Tertiary (Beltz, 1944). The Oligocene was a period of marine transgression which reached its maximum extent during the Middle Miocene. Tertiary sedimentation appeared to have started with the development of isolated sub-basins in which sands, conglomerates, and carbonaceous shales were laid down. These sub-basins merged into the Aceh basin during the Lower Miocene when subsidence became less pronounced. During this period of quiescene there was widespread carbonate and arenaceous deposition. In the central part of the basin great thicknesses of the calcareous shales of the Peutu Formation were deposited containing reef complexes such as the Arun Limestone. These are the source of the most important hydrocarbon occurrence in the Aceh basin, viz., the Arun Gas Field, which has recoverable gas reserves equivalent to about three billion barrels of oil. It is believed that the Arun field constitutes 85 per cent of the basin's hydrocarbon accumulations (Kingston, 1978).

Renewed subsidence occurred during the Middle Miocene over the entire basin area, at a rate of about 300 m per million years in the central part, where some 5 500 m of sediments were laid down. During this period thick deposits of shales of the Baong Formation buried the pre-Baong sediments, and were in turn overlain during the Upper Miocene/Pliocene by the reservoir sands of the Keutapang and Seurula Formations. The regressive, mainly sandstone and shale sequence of these Formations was derived partly from the clastic materials of the uplifted Barisan Mountains and partly from the foreland (Pulunggono, 1976; Kingston, 1978). In-filling of the basin ended in the late Pliocene, and the basin was later subjected to Plio-Pleistocene folding, producing the present structural pattern (Pulunggono, 1976).

The Baong and post-Baong sands are the source of the other 15 per cent of the hydrocarbon accumulations of the Aceh basin. The reservoirs are found near the eastern edge of the basin or on the basin foreland and are small and scattered. Thirty-seven oil accumulations have been found at depths of about 1 000 m. Most of them are now depleted, and only twelve fields continue to produce the very high gravity oil characteristic of the basin (Kingston, 1978).

THE CENTRAL SUMATRA BASIN (NO. 2 IN FIG. 3.6)

The basin shares a similar geological history with the other foreland basin of South Sumatra. The marine transgression of the Lower Oligocene was accompanied by the deposition of the basal sands of the Sihapas Group, the source beds of most of the oil found in this basin. In the Middle and Late Miocene, when the transgressive period closed and the Barisan orogeny began, the marine shales that were deposited provided cap rocks for the rich reservoir rocks of the Lower and Middle Miocene.

The Central Sumatra basin has formations 2 350 to 3 500 m thick, and is separated from the North Sumatra (Aceh) basin by the low Asahan arch. The basin has an area of about 52 000 sq. km, and is bounded on the south-west by faults and uplifted exposures of the pre-Tertiary rocks in front of the Barisan Range, and on the north-east by the Sunda Shelf. In contrast to North and South Sumatra, no oil was produced in Central Sumatra until 1938 as exploration efforts were initially confined to the upper regressive facies, which in Central Sumatra have proved to be non-productive. However, drilling in the lower transgressive series led to the discovery of the most prolific field of this region—the Minas field (Fig. 3.6). Over 250 wells have been drilled in this field, which is about 24 km long and 8 km wide. Minas has already produced over 2 billion barrels of oil. Production is from Lower Miocene sands at a depth of about 750 m. The Duri field, which has produced 300 million barrels of oil, lies adjacent to the Minas field.

THE SOUTH SUMATRA BASIN (NO. 3 IN FIG. 3.6)

This basin is separated from the Central Sumatra basin by a low basement arch, but has a similar geological history of an early Tertiary regression followed by an Oligocene to Late Miocene marine transgressive period. During the earlier part of this transgressive phase, the basal Telissa sands were laid down, while during the early Middle Miocene shallow shelf carbonates (the Baturaja Limestone Formation) were deposited. The Late Miocene saw the filling of the basin with marine shales. Structurally, the basin is made up of two sub-basins—the Jambi and the Central Palembang. The most conspicuous structural features of both the Central and South Sumatra basins are the folds and faults formed during the Plio-Pleistocene period of orogenic activity, when the collision of the

Indian Ocean plate against the Sumatran portion of the South-East Asian plate is postulated to have brought about the final uplift of the Barisan Mountains (Pulunggono, 1976).

The South Sumatra foreland basin has a maximum total sedimentary thickness of about 5 000 m. The basin is about 78 000 sq. km in area, and is bounded on the north-east by the Sunda Shelf and on the south-west by the Barisan Range. The production history of this basin dates back to the end of the nineteenth century, and a detailed picture of its geological history has been built up over the years through extensive exploration and surveys, with some wells penetrating the entire Tertiary section. As Figures 3.2 and 3.3 show, oil is found in the upper regressive series located nearer the Barisan Range as well as in the lower transgressive series in the shallower part of the basin near the Sunda Shelf. However, most of the fields in the upper regressive series are now depleted, and production is mainly from the lower transgressive facies, in the prolific sands of the Lower Telissa Formation. The Baturaja Limestone is known to be a prolific reservoir for oil and gas. The most important oil field is the Talang Akar−Pendopo field covering an area 19 km long by 3 km wide. The productive zone of the Telissa sands has a combined gross thickness of about 230 m. Yields from this field are about 50,000 barrels per day. The oils from these structural closures have a heavy paraffinic base.

NORTH-WEST JAVA BASIN (NO. 4 IN FIG. 3.6)

The north-west Java foreland basin area, comprising the West Java, Sunda and Billiton basins, has recently been proved to contain substantial hydrocarbon accumulations offshore in the Java Sea. Current total production from this basin is about 200,000 barrels of oil per day. The West Java basin itself has three sub-basins, the Jatibarang, Ciputat and Pasir Putih (Soejitno & Yahya, 1974). The entire North-West Java basin, onshore and offshore, is now regarded as a major producing area in Indonesia. The basin consists of a thick (6 000 to 8 000 m) section of Tertiary sediments near the axial belt which thins with distance northwards. The hydrocarbon producing zones in the North-West Java basin are usually confined to sedimentary sections of Eocene and Oligocene age. However, the oil and gas produced onshore from the Jatibarang Volcanic Formation are from fractured tuffs.

The offshore extension of the North-West Java basin has only about 3 000 m of typical marine to paralic shelf facies of clays, marls and reef limestones intercalated with sandstones. Drilling in 1969 resulted in a discovery of oil in Miocene marine sandstones in the Arjuna area. New offshore fields have been discovered south-east of Sumatra and north-west of the West Java basin. The Cinta field, tapping oil from an Early Miocene limestone at 900 m depth and sandstone at 1 100 m depth, began commercial production in September 1971 and by March 1972 was producing about 44,000 barrels per day. Three other fields—Kitty, Gita, and Zelda—were discovered on the same trend as the Cinta.

During the Paleogene transgression (Eocene to Early Oligocene) the Sunda and Arjuna basins were filled with non-marine sediments derived from granites of the Sunda Shield. The major Neogene transgression, which started in Late Oligocene time and continued until the Middle Miocene resulted in rapid basin fill-up sedimentation. The Arjuna and Sunda basins received transgressive sandstones near the Sunda Shield, but these were thin at the basin centres (Beddoes, 1980). Pulunggono (1976) has postulated three sedimentary environments during Late Oligocene and Early Miocene times in the West Java basin—shelf, shallow marine, and possibly deep marine. The Early Miocene also witnessed carbonate deposition (Baturaja Formation) in the Sunda platform area and calcareous shale deposition in the basin areas south of the Sunda Shield, including the Western Java Sea basins. In these basins, coarse clastics of the Upper Cibulakan Formation were laid down over the Baturaja Formation during the Middle Miocene uplift of the Sunda Shield. During the late Middle Miocene transgression which followed the basins received clays and marls (Beddoes, 1980). The Tertiary sediments which blanket the North-West Java basin area are up to 4 500 m thick.

Although surface oil seeps occur in this basin and oil has been produced onshore for many years, such production was on a small scale. It was only in the 1970s that the high potential for hydrocarbon accumulations in the offshore part of the basin became apparent, with yields obtained from a large variety of rocks, including basement rocks, Oligocene volcanics, Oligocene channel sands, Oligocene–Miocene carbonates, Mio-Pliocene sands, Pliocene carbonates and Plio-Pleistocene sands. The multiplicity of stratigraphic targets, added to the variety of source beds, point to the excellent prospects for oil and gas accumulations in this basin. The recent discovery of a major oil field in Oligocene-Miocene carbonates at Rama in the Sunda sub-basin has increased the potential of the North-West Java basin by the inclusion of carbonate reservoirs as possible areas of hydrocarbon accumulations.

EAST JAVA BASIN (NO. 5 IN FIG. 3.6)

This foreland basin shares a common boundary with the West Java basin along the Karimunjawa and Bawean arch, while the north shelf of the basin runs along the southern limits of the East Kalimantan basin. The total area of the basin is about 18 000 sq. km. It shares a similar geological history with the West Java foreland basin discussed earlier. During the Paleogene transgression, Eocene marine sediments were laid down on the eastern and southern parts of the basin. The northern flank of the Madura Trough and the eastern edge of the Java Sea were covered by shelf carbonates. The Paleogene transgression came to a close at the end of Early Oligocene time, followed by a regression which lasted until the Late Oligocene. The major Neogene transgression then commenced, and the basins on the western edge of the East Java Sea were covered by coarse clastics, but those on the east received thick deposits of calcareous shale and shelf carbonates. During the latter part of this transgressive period patch and pinnacle reefs grew within the platform and basinal areas.

At the end of this transgression (mid-Miocene time) an uplift of the Sunda Shield brought on a brief regressive phase, when fine-grained clastics were deposited in the basin. A late Middle Miocene transgression then occurred, followed by a Late Miocene to Pliocene regression, during which the entire basinal area was covered by fine-grained clastics of the Cisubuh and 'GL' Formations (Beddoes, 1980) (Fig. 3.8).

The known Tertiary sequence of the East Java basin is over 7 600 m thick (Pulunggono, 1976). On the island of Madura, some 7 000 m of

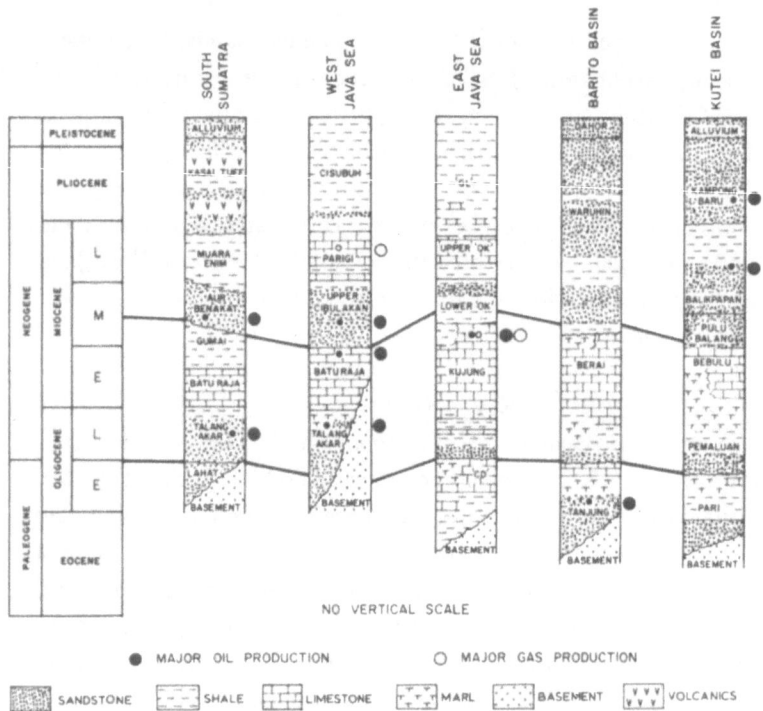

Fig. 3.8. Generalized stratigraphic sections of five Tertiary basins in
Indonesia peripheral to the Sunda Shield
Source: After Beddoes, 1980.

Miocene and Oligocene beds have been penetrated by the drill. Since
1887, the oil fields discovered in this basin have been small
accumulations associated with faulting and/or stratigraphic traps in
the upper regressive series of Upper Miocene and Lower Pliocene
sands. The largest field was Kawengan, discovered in 1926 from
drilling in upper Miocene sands at depths of 300 to 700 m. In 1972, a
small offshore oil find was made by Cities Service in the East Java Sea,
and was put into production in November 1975. This field, named
Poleng, is located 24 km offshore of Madura island in 55 m of water. It
had produced over 2 million barrels of oil by mid-1978, but was
subsequently shut-in due to its increasingly high production of gas
with diminishing oil flows.

Most of the fields in this basin have been producing for many years, but are of marginal importance today. However, future oil possibilities may exist in the lower transgressive formations in the offshore areas north of the basin, in the direction of the Sunda Shelf. Carbonate platform reefoid deposits and pinnacle reefs hold the greatest exploration promise, and have recently become a prime objective in the search for oil in the East Java—Madura basin (see Soeparjadi et al., 1975). A number of wells have been drilled in this basin area, and in mid-1977 oil was tested from an Eocene carbonate reservoir in the East Java Sea. Although reservoirs, seals and structures are present in this basin no major fields have yet been discovered in the area (Beddoes, 1980).

THE KALIMANTAN FORELAND BASINS

These basins occupy a foreland depression on the eastern edge of the Malaysian craton. The depositional history of the Kalimantan foreland Tertiary basins began as early as the Eocene and ended in the Pleistocene during which period over 12 000 m of sediments were laid down in the thickest part, as in the Kutei-Melawi basin (Rose & Hartono, 1978). During the Eocene, the basins peripheral to the Sunda Shield, including the Barito, Kutei and Tarakan basins of East Kalimantan, received fine-grained clastics derived from marine environments in their central parts and non-marine sands and conglomerates on their flanks. In the Barito basin the first Tertiary sediments, the Tanjung Formation (Eocene age), were laid down over a fairly low relief pre-Tertiary basement (Sinegar & Sunaryo, 1980).

From the Eocene to Early Oligocene these basins received shelf limestone on the flanks and bathyal shales in their central parts as the seas slowly transgressed the region. At the end of this period Kalimantan was uplifted, causing a brief regression and deposition of coarse-grained clastics. The major Neogene transgression which followed provided the Kutei and Tarakan basins with fine-grained clastics, while the platform areas of the Paternoster Shelf, Mangkalihat Peninsula, and the northern portion of the Tarakan basin experienced stable marine conditions with resultant deposition of shelf limestone, patch reefs and shelf-edge reefing around the platform margins (Beddoes, 1980). In the Barito basin the shelf limestone of the Berai Formation was laid down during this period, with extensive local growth westwards from the Paternoster Shelf and southwards to the Java Sea (Rose & Hartono, 1978; Sinegar & Sunaryo, 1980).

The Neogene transgression gave way to a major regression in the mid-Miocene when tectonic forces uplifted Central Kalimantan and, in the south, formed the Meratus Range dividing the Barito from the Kutei basins. In the Barito basin the emergence of the Meratus Range was accompanied by a rapid deepening of the basin. Erosional material (coarse clastics) contributed to the thick paralic Warukin Formation of this basin, while finer-grained clastics were carried south into the Java Sea. In the Kutei and Tarakan basins eastward flowing rivers—the ancestral Mahakam River in the Kutei basin and the ancestral Sesayap in the Tarakan basin—initiated the deposition of massive deltaic complexes. During this regression, the ancestral Mahakam River deposited thick wedges of the Balikpapan Formation sandstone and shale in the Kutei basin. In the Tarakan basin, the Central Kalimantan uplift brought about a change from open marine to paralic conditions, and large quantities of clastic sediments derived from the emergent areas were deposited by the ancestral Sesayap River to form constructive deltas which prograded from west to east. These sediments formed the Meliat Formation (see Samuel, 1980).

A Late Miocene transgression deposited extensive carbonates over the deltas, but this transgression was short-lived, ending when a major orogeny occurred during Late Miocene–Pliocene time. That orogeny uplifted Central Kalimantan, the Meratus Range and the Mangkalihat Peninsula, and provided abundant material for all the Kalimantan basins during the regression which followed and continued into the Pleistocene (Beddoes, 1980). In the Barito basin the sediments of the Dahon Formation were laid down during this period, as were the deltaic deposits of the Kampong Baru Formation in the Kutei basin and the Tarakan Formation in the Tarakan basin.

THE BARITO BASIN (NO. 6 IN FIG. 3.6)

This covers an area of 70 000 sq. km and occupies the south-western part of the Kalimantan basinal area. It is bounded on the west by the Sunda Shelf and on the east by the Meratus Range. The Paternoster High separates the basin from the Kutei basin to the north. The southern limit of the basin extends into the Java Sea. The Tertiary deposits on the west are only a few hundred metres thick, but thicken to about 6 000 m along the front of the Meratus Range near Tanjung. The strata range in age from Eocene to Pleistocene, and are steeply upturned in a narrow belt along the Meratus Range and the Kasale Range.

Oil seeps from the Eocene and Miocene beds occur in this area. Exploration in this basin area before the Second World War in the Pacific led to the discovery of the Tanjung field. The source rocks are believed to be the neritic shales and marls of the Tanjung Formation, while the reservoir rocks are the deltaic sandstones of the same Formation. Sinegar and Sunaryo (1980) believe the hydrocarbon potential of the Tanjung Formation to be high because the three factors essential for hydrocarbon accumulation—source rocks, reservoir rocks and traps—exist in it. The source rocks are expected to be widely distributed, particularly in the South Barito Deep. Reservoir rocks of transgressive and deltaic coarse clastics are also expected to occur widely in the undrilled area, with traps being elongated anticlines similar to the Tanjung field anticline and Kambitin structure.

THE KUTEI BASIN (NO. 7 IN FIG. 3.6)

This is the largest of the Kalimantan basins. It is bounded on the north by the Mangkalihat Peninsula, on the east by the trough of the Makassar Strait, on the south by the Paternoster Cross High, and on the west by the high relief of Central Kalimantan. The basin consists of two large present-day river basins—the Kutei and the Mahakam—and their offshore extensions. Tertiary clastic rocks up to 12 200 m thick are found in the basin.

Folding appears to have been caused by gravitational compression rather than by tectonic forces, as attested by the juxtapositioning of tightly folded onshore anticlines such as Sanga-Sanga with very gently folded offshore structures such as Attaka and Bekapai (Rose & Hartono, 1978).

The upper Mahakam basin is underlain by gently folded and

faulted sandstones, shales and limestones of mainly marine origin and Eocene and Oligocene in age. These deposits are from 7 600 to 11 600 m thick. The section in the large central Mahakam basin consists of marine deposits interbedded with fresh or brackish-water sediments. The exposed beds on the basin margins are steeply folded. The lower Mahakam basin is underlain by steeply folded strata of Miocene marine sediments interbedded with Pliocene terrestrial sediments. The oil fields in this area are located near or along the east coast, in Middle Miocene regressive sands on tightly folded anticlines. The oil accumulations of the large Sanga-Sanga field are in the Middle Miocene sands of the Balikpapan Formation.

The inner section of the Kutei basin forms a northern extension of the upper Mahakam basin. The coastal section includes the Sangkulirang Bay region and the Kutei district south of the Bay. The average section in this basin includes nearly 4 000 m of Upper Tertiary sediments (with limestones dominating the Oligocene) and over 9 000 m of Lower Tertiary sediments (with sandstones and shales dominating the Eocene). Oil seeps occur in the Upper Miocene and Pliocene sediments in the coastal region. The Attaka, Badak, Handil, Bekapai and other recent discoveries in the Kutei basin are all associated with the Upper Miocene and Pliocene delta front deposits of the ancient Mahakam River.

Since 1970, when the first offshore oil was discovered (Attaka field) 35 km off the Mahakam River delta, the offshore Kutei basin has been proved to be a prolific hydrocarbon province. Oil is found in Neogene regressive sandstones, in deltaic reservoirs classified as play-type No. 3 by Soeparjadi et al. (1975) and as play-type 'D' by Beddoes (1980). Indonesia's second largest oil field, Handil, was discovered here in 1974, in the Mahakam River delta. Handil, covering an area of 40 sq. km, has oil in 150 sand reservoirs of Middle to Late Miocene age, lying at depths of between 500 and 2 900 m (Sauphanor & Seguin, 1980). The Attaka field, which has a longer production history, has recorded a cumulative output of 228 million barrels up to mid-1979. The giant-size Badak gas field is also in this basin.

THE TARAKAN BASIN (NO. 8 IN FIG. 3.6)

The Tarakan Basin is a narrow basin covering an area of 52 000 sq.

km. It is separated from the Kutei basin by an area of uplifted rocks forming the Mangkalihat Peninsula. The Semporna High forms its northern boundary, while the pre-Tertiary rocks of the Kuching High set its western limits. The basin extends offshore to the Makassar trough. Samuel (1980) has divided the Tarakan basin into four sub-basins, of which the Tarakan sub-basin has the youngest depositional history. Beginning with a marine transgression in Oligocene to Early Miocene time, when the marine clays, shales, marl and limestone of the Naintupo Formation were laid down, the sub-basin underwent a period of uplift in late Early Miocene to Middle Miocene, which resulted in large volumes of clastic material being deposited to form deltas prograding from west to east. During the Plio-Pleistocene the ancestral Sesayap River contributed large quantities of sandstones, shales and coals of the Tarakan—Bunyu Formation. The sub-basin was uplifted, folded and faulted in a tectonic event in the Late Pliocene or Early Pleistocene. The present-day Tarakan and Bunyu islands were the result of a Pleistocene eustatic rise in sea-level which moved the Kalimantan shoreline west.

Up to recently, oil production was confined to the Tarakan and Bunyu islands, in regressive clastic reservoirs. Over 200 million barrels of oil have been produced in Tarakan since 1906. Exploration onshore resulted in the Sembakung oil discovery in 1976.

THE SALAWATI BASIN (NO. 9 IN FIG. 3.6)

Western Irian Java, unlike western Indonesia which was subjected to tectonics and differential movements resulting from the collision of a bordering cratonic and a continental plate, lies in a complex area of interaction of three major crustal plates: the Pacific Oceanic plate to the north, the Australian continental plate to the south-east, and the Eurasian continental plate to the south-west (Froidevaux, 1977).

Within western Irian Jaya is a geologic province consisting of the Vogelkop Peninsula and the associated southern shelf. The boundaries of the province are the Lengguru fold belt to the east, the Seram Sea to the south and the Sorong fault block to the north, with the western margin near Halmahera. A downwarp of the eastern and western margins during the Middle Miocene resulted in the formation of the Bintuni and Salawati basins (Pulunggono, 1976).

The depositional history of the geological province during the Tertiary consisted of two complete cycles of transgressions and regressions: Eocene transgression, Oligocene regression, Middle Miocene transgression and Late Miocene regression. In the Salawati basin widespread marine transgression during the Middle Miocene initiated a phase of carbonate deposition and reef development. This was a period of little tectonic activity. Regional standstill of the sea was accompanied by the deposition of carbonates (Kais Formation) on shallow shelf areas, building a platform out into the basin over the basinal limestones of the Klamogun Formation. Pinnacles reefs 300 to 750 m thick grew on this platform. The regional sea at that time extended northwards beyond the present-day Sorong fault zone (Froidevaux, 1977).

Areas in which there was no reefal activity were filled by calcareous shales of the Klasafet Formation, which eventually buried the pinnacle reefs and provided top and side seals. A regressive phase was established at the end of the Miocene, following uplift of the Australian craton. Reef build-up came to an end and the basin's deposits subsequently consisted of Pliocene, Pleistocene and Recent shale and clastics. Plio-Pleistocene orogenic movements resulted in the formation of the Sorong fault, which marks the basin's northern boundary, while to the east the Lenguru belt was folded and uplifted. The tectonic activities have also uplifted, faulted, and tilted many of the pinnacle reefs (Soeparjadi, et al., 1975; Pulunggono, 1976).

Oil was discovered in the Salawati basin just after the war, and two fields—Klamono and Wasian/Mogoi—were producing 4,000 barrels per day in 1948. In recent years exploration centred on the Middle to Late Miocene oil-proved pinnacle reefs. Several discoveries have been made, of which the Walio oil field is the largest. The reef reservoirs have produced nearly 80,000 barrels per day, with individual wells yielding the highest flows in Indonesia. Production comes from Late Miocene Kais Formation in stratigraphic or combination fault/stratigraphic traps. Source rocks are mainly from adjacent organic-rich marine shales.

While the pinnacle reefs have excellent reservoir quality, their reservoir area is commonly small, ranging from 240 to 400 ha. Their hydrocarbon columns may be as thick as 130 m (Vincelette, 1973). Although the reefs are small and scattered throughout the Salawati basin, their hydrocarbon potential is regarded as very good, and they have become a prime target of current exploration activity.

THE WAROPEN BASIN (NO. 10 IN FIG. 3.6)

This other foreland basin in Irian Jaya differs from all the other foreland basins of Indonesia in that it is still non-producing and little explored, although believed to hold significant oil potential. The Tertiary section of the Waropen basin is up to 9 000 m thick, and is made up of transgressive basal clastics of Eocene and Oligocene age overlain by a marine transgressive section of fine clastics interbedded with shallow water limestones with reefal build-ups of Lower Miocene age. Uplift of the Central Highlands in the Middle Miocene was followed by deposition of coarse clastics for the remainder of the Tertiary. These clastics, which are up to 2 000 m thick, are apparently good reservoir rocks. The existence of large anticlinal folds points towards the possibilities of large hydrocarbon accumulations. Further promising exploration targets are Lower Miocene carbonates, including reefs, the equivalent of which has proved so prolific in the Salawati basin. The Mamberamo block covering 14,675 sq. km of the basin has been contracted out to Shell on a production-sharing agreement (*Petroleum News*, January 1980).

Cratonic Basins

Cratonic basins are formed when stresses cause foundering or downwarping within a craton. The Indonesian cratonic basins appear to have been the result of tension-induced foundering along lines of weakness. There are three such basins: the Arafura basin (No. 13 in Fig. 3.6) on the northern shelf of the Australian craton, and the Natuna and Melawi basins on the Malaysian craton (Nos. 11 and 12 in Fig. 3.6). The Arafura and Natuna basins are very large, while the Melawi basin in Kalimantan is small and as yet non-producing.

The *Natuna Basin* extends beyond Indonesian territory to cover Sarawak, Sabah and Brunei. The Indonesian part of the basin is wholly offshore and, unlike the Malaysian and Brunei section, has only recently started to produce oil. Recent geological data indicate that the Natuna basin is made up of two sub-basins: the West Natuna basin and the East Natuna basin, separated from each other by the north/south trending Natuna ridge. The East Natuna basin was formed in Eocene time when the South China Sea oceanic plate subducted under Borneo along the Lupar subduction zone. This was accompanied by volcanic activity and emplacement of granites along the edge of the craton, including the island of Natuna. Shallow and deep water facies were deposited in the basin during the Oligocene, after the unconformity related to crustal uplift along the Natuna batholithic arc subsided. Basinal in-filling continued to the Pleistocene, with the Pliocene witnessing progradation of the basin when deposition was mainly in synclinal lows and at the shelf edge.

The West Natuna basin, part of a larger basin developed in the Gulf of Thailand and the Sunda Shelf, was formed during the Oligocene when the Lupar subduction zone became inactive. The cause of basin formation has been ascribed to the foundering of the Sunda landmass. The depositional environment was non-marine to brackish. During this period, reservoir sandstones of the Gabus Sandstone unit derived from erosion of the granitic and metamorphic basement were laid down. These were overlain by shales of the Barat unit, believed to be a hydrocarbon source and seal. A marine environment prevailed in the basin during mid-Miocene time, when the sill on the Natuna batholithic arc was breached. During the Pliocene the basin was buried by open marine muds (White & Wing, 1978).

Exploration for hydrocarbon accumulations in the Natuna basin started in late 1968 and resulted in the discovery of the Udang field in the West Natuna basin by the CONOCO group.

The hydrocarbon potential of the Natuna basin appears to be good, as attested by the fact that a large gas field has also been discovered offshore near Natuna Island (*Petroleum Economist*, May 1980).

The *Arafura Basin*, located on the Australian craton, contains the following Tertiary sequence: Early Tertiary coarse clastics; Oligocene and Miocene shallow water limestones; transgressive fine clastics; Pliocene and Pleistocene regressive shales and coarse clastics. The combination of carbonate reservoir rocks with shale source rocks appears to point towards a good hydrocarbon potential. But again, as in the case of the Natuna basin, the absence of structures (except immediately adjacent to the Central Highlands uplift), as well as the low geothermal maturation level of the rocks, serves to depress this potential. Although exploration in the Arafura basin is now in progress, no significant hydrocarbon accumulation has yet been discovered.

Outer Arc Basins

There are four outer arc basins in Indonesia: the Sumatra Outer Arc basin (No. 14 in Fig. 3.6.), the Java Outer Arc basin (No. 15), the Lombok basin (No. 16) and the Savu basin (No. 17). These basins were formed on the boundaries of colliding plates, adjacent to subduction zones. The Sumatra and Java Outer Arc basins, for example, were formed on the boundary where the ancestral Indian Ocean plate collided against the Malaysian craton, creating the Sumatra-Java subduction zone. The sediments in the basins are mainly marl and limestone interbedded with pyroclastics and volcanic erosion products. The paucity of source rocks makes such basins low in hydrocarbon potential. The Sumatra basin, for instance, has an Eocene to Recent sedimentary section consisting of shelf limestones and fine clastics interbedded with volcanic-clastics, but no organic-rich shales. Although there are some indications of oil and gas in the northern part of the long and narrow basin no commercially important hydrocarbon accumulations have yet been found.

Inner Arc Basins

Fletcher and Soeparjadi (1976) use this term to include not only the classic inner arc basin, but also other basins formed in Indonesia as a result of tensional stresses or compressional forces generated through the collision of the Eurasian, Pacific and Australian crustal plates. Within the triple junction of these plates, and covering much of central Indonesia are eleven inner arc basins (Nos. 18 to 28 in Fig. 3.6). Unlike the other Indonesian basins, these were formed in the

Late Tertiary and have only a thin cover of sediments, mainly outer shelf, deep water facies with poor reservoir characteristics. A high percentage of the sediments are thermally immature, and this combined with the lack of large structures of tectonic origin, would point towards a poor hydrocarbon potential.

However, some of these inner arc basins do contain indications of hydrocarbons. The North Seram basin (No. 22 in Fig. 3.6.) has a small oil accumulation. Seram itself is an island of complex geology, with metamorphic and igneous rocks and Jurassic and Triassic sediments. Late Tertiary and Pleistocene sediments occur on the north coast and its offshore extension. The Bula oil field, on the easternmost embayment, was the only offshore field developed before the Second World War. Production was mainly from individual Pleistocene bar sands, 10 to 20 m thick, lying 100 to 300 m in depth. About 7.5 million barrels of oil were recovered from the field up to the end to 1940, when production ceased. The field was re-opened in 1970, with a maximum daily output reaching 2,000 barrels in mid-1971, but declining to 927 barrels/day in mid-1979 *(Oil & Gas Journal,* 31 December 1979). Oil prospects in the Plio-Pleistocene beach, bar and turbidite sands and reef limestones appear good and 20 million barrels of oil are estimated to be reservoired in these sands and reefs (Gribi, 1973; Soeparjadi *et al.*, 1975).

IV

Exploration

OIL exploration in an unexplored area is both a time-consuming and an expensive operation. Different strategies are followed on land and at sea, with relative costs varying at each stage of exploration. As a general rule the cost of a general survey is higher on land than at sea, but the cost of well drilling is substantially higher at sea than on land. In the Indonesian context where some 13,000 islands are spread out across 5 000 km of sea and where hydrocarbons are found onshore as well as offshore, exploration costs can vary quite considerably with local conditions of terrain, vegetation, water depth, distance from shore, accessibility and remoteness of site.

The wide variety of conditions in Indonesia has in turn necessitated the use of a wide range of exploratory techniques in the search for oil and gas accumulations. Exploration in an unexplored area starts off with a general survey which seeks to provide an overview of the area. Such a survey can cover large expanses of territory at relatively low cost. Photogeology, using aircraft, is employed in those onshore areas where the surface of the earth's crust provides an indication of the geology of deeper formations. Aeromagnetic surveys are used for onshore as well as offshore exploration. Figure 4.1 shows the annual extent of aeromagnetic surveys carried out in Indonesia between 1966 and 1979. Altogether a total of 172 000 km lines was recorded. Most of such surveys conducted by companies operating under production-sharing contracts were completed by 1975.

The related phase of exploration, initiated immediately after some general knowledge has been gained about the area, consists of intensive surveys to pin-point potential oil-bearing basins. On land, more specific information can be obtained by surface geological

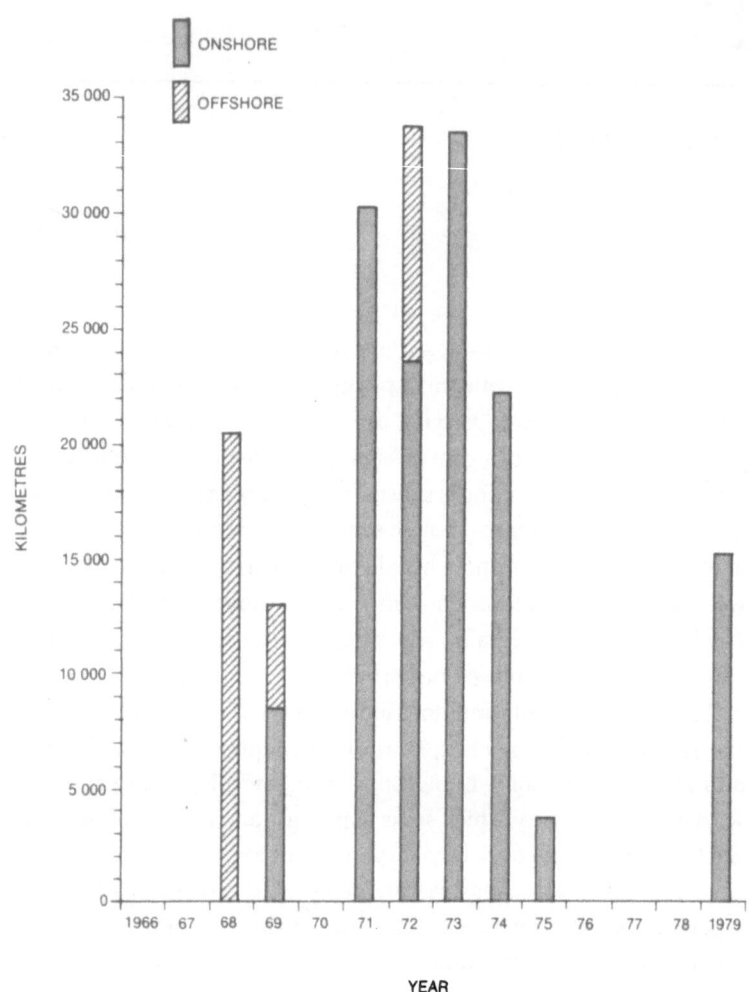

Fig. 4.1   Aeromagnetic surveys conducted in Indonesia, 1966–1979
Source: IPA Professional Division, 1980.

mapping using field crews. In the rainforest areas of Indonesia such mapping is an often slow and expensive undertaking. Figure 4.2 shows the pattern of such investigations conducted during the 1966–79 period. It will be seen that most of the surveys were under-taken between 1971 and 1975, mainly by companies operating under

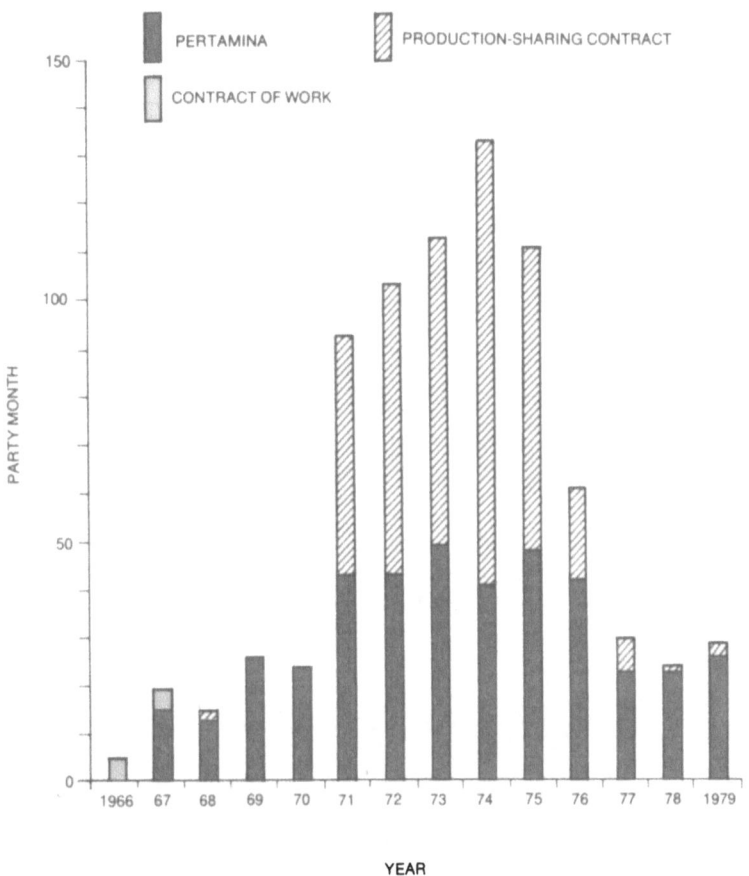

Fig. 4.2 Surface geology surveys conducted in Indonesia, 1966–1979

production-sharing contracts, and that, as in the case of aeromagnetic surveys, surface geological surveys tapered off after 1975. The surveys were conducted by PERTAMINA and the production-sharing contractors, with the Contract of Work companies involved only in 1966 and 1967.

Where there are indications of the existence of promising structural traps, a survey by seismic reflection may be undertaken. Such seismic surveys will indicate or confirm the locations for drilling exploration wells. Again costs are high in Indonesia because of difficult rainforest

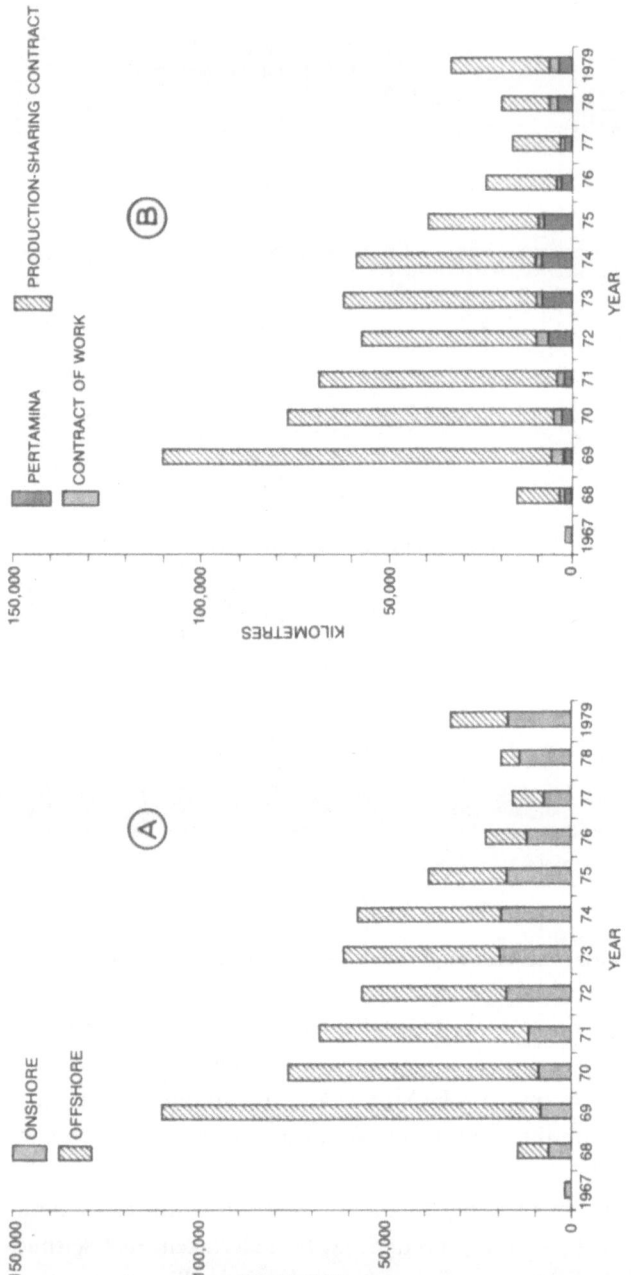

Fig. 4.3 Seismic surveys conducted in Indonesia, 1966—1979
(A) onshore and offshore
(B) by types of contract

Source: IPA Professional Division, 1980.

conditions and the need to have multiple coverage in order to compensate for the interference created by the weathered zone above the bedrock. For example, the Elf-Aquitaine group incurred costs of $15,000 per mile for seismic operations in rainforest areas in Indonesia (in 1975) as compared with $2,000 per mile in less difficult country. Three-quarters of the total cost of seismic operations were incurred for onshore work, which however, produced only 16 per cent of the total mileage covered (Mainguy, 1976).

In contrast, seismic operations in offshore areas cost only between $140 and $540 per mile, although shallow water seismic work in swamps and deltas may cost more. Such seismic measurements, involving towing a line of geophones behind a ship, contribute considerably to the discovery of potential structural traps. Figure 4.3 provides details of the extent of seismic surveys conducted onshore and offshore in Indonesia, as well as their distribution according to the types of operating companies, for the period 1966–79. The most significant feature revealed by these figures is the fact that nearly 90 per cent of the 573 500 km surveyed were conducted by production-sharing contract companies which operate largely in offshore areas. Most of the surveys were carried out between 1969 and 1974, with 110 000 km recorded during the single year 1969.

The search for oil in Indonesia, fuelled by high success ratios, has intensified as the number of oil companies operating in the country increased rapidly in the late 1960s and early 1970s. Such a competitive situation has contributed to widening the range and diversity of exploration techniques employed in the location of new hydrocarbon deposits. On a macro-scale, Earth Resources Technology Satellite (ERTS) imagery is being used to delineate regional lineament trends and large tectonic features. Side-looking Aerial Radar (SLAR) has proved highly useful in Kalimantan, Sumatra and Irian Jaya in exposing geological structures hidden by cloud cover and jungle. Aerial photography using colour, infrared or black and white film has helped exploration teams in various ways, particularly in the location of structural anomalies.

The international oil companies have been using digital seismic methods offshore and onshore since 1965. New seismic data acquisition techniques and modern high resolution seismic data processing are enabling the interpreter to identify not only the simple structural traps but also more complex targets such as reefoid anomalies within thick carbonate sequences. Advances in prospecting techniques now allow the explorationist to predict fluid content as well as the entrapment structure (Stommel & Graul, 1978). Field seismic processing units have been used for on-site digital data processing by Trend Exploration in their contract area in the remote jungle areas of Irian Jaya to identify the North Kasim anomaly, have it detailed by seismic survey, and have the first well spudded within three months of the find (Soeparjadi *et al.*, 1975). Seismic processing techniques have been used with equal effectiveness in offshore areas to detect reflection amplitude anomalies ('bright spots').

No final proof of the existence (or absence) of hydrocarbons in an area can be obtained except by the drilling of wells. This represents the final phase in exploration work. The drilling of an exploration well (wildcat well) in a new area is expensive. The costs are multiplied several times for offshore drilling. Mainguy (1976) quotes the following 1974 figures: the daily rental of a drilling installation capable of drilling to a depth of 3 500 m was $5,000. The rental for an offshore drilling unit varied from $12,000 for a unit mounted on swamp barges to $40,000 per day for an ultra-modern self-contained dynamic positioning unit capable of drilling in more than 200 m of water.

In the first phase of exploration drilling the search for oil is centred on locating structural traps, the existence of which has been indicated by detailed geologic or seismic surveys. After sufficient drilling has been completed to provide a general stratigraphic picture in the geologic province, the search is extended to cover stratigraphic and reef traps such as buried reefs, sand lenses and 'wedge-outs'. Such traps may be of considerable importance. Most of the oil discovered in Indonesia has been from structural, mainly anticlinal traps. But more recently attention is also being directed to stratigraphic and reef traps. Significant discoveries have been made in the Salawati basin where pinnacle reefs produce over 80,000 barrels per day of high gravity, low pour-point oil. At Rama in the Sunda sub-basin of North-West Java about 30,000 barrels of oil per day are being produced from reefal carbonates of Middle Miocene age. The potential for further discoveries in stratigraphic pinchouts and

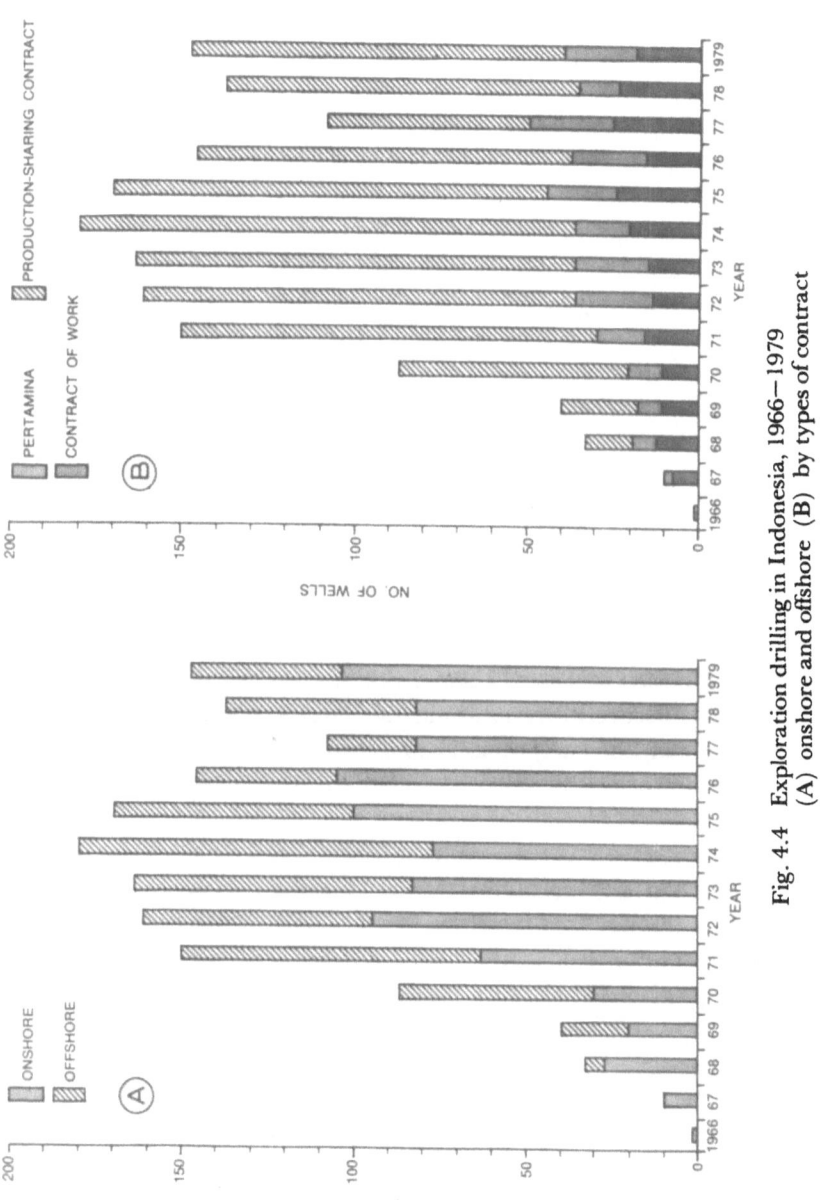

Fig. 4.4 Exploration drilling in Indonesia, 1966—1979 (A) onshore and offshore (B) by types of contract

Source: IPA Professional Division, 1980.

carbonate sequences in the Sumatran and Irian Java foreland basins, in North-West Java and in Kalimantan is now regarded as good (Fletcher & Soeparjadi, 1976).

It is exceptional for the first wildcat well drilled to be successful. Such 'dry' wildcats will, nevertheless, provide information which may eventually prove to be highly important. The fact that a number of wells will usually have to be drilled in an area before a commercial deposit of oil or gas is discovered is a well-known risk in the oil industry. Exploration costs are consequently increased. To cover the risk oil companies usually spread their exploration efforts as widely as possible. It is axiomatic that to remain viable the companies would have to discover oil or gas in sufficient quantities not only to pay for the unsuccessful ventures, but also to provide adequate returns to capital invested.

Figure 4.4 shows details of exploration drilling conducted in Indonesia between 1966 and 1979. The number of wells drilled increased dramatically from only one in 1966 to 180 in 1974. Altogether 1,536 wells were drilled during this 1966–79 period, of which 73 per cent was by production-sharing contractors, 14 per cent by Contract of Work companies, and 13 per cent by PERTAMINA. Forty-three per cent of the wells were drilled in offshore areas, with oil and gas discoveries recorded in the Java Sea, the sea off East Kalimantan, and lately in the Natuna Sea.

The ratio of success wells to the number of exploration wells drilled

TABLE 4.1

SUCCESS RATIOS* IN INDONESIA, 1973–1977

Company	1973	1974	1975	1976	1977
PERTAMINA (onshore)	25	40	40	36	60
Contractors (onshore & offshore)	16	29	40	29	40
Average	20.5	34.5	40	32.5	50

Source: Pulunggono (1979).

* $\frac{\text{Total successful wells}}{\text{Total exploration wells}}$ (%)

is termed the success rate. A successful or productive well is usually defined as one 'productive of some oil or gas'. The success ratio in the U.S., for example, has averaged about 10 per cent (one productive well out of ten drilled) since 1945 (Mainguy, 1976). In comparison, the success ratios for Indonesia are much higher, as Table 4.1 shows. In computing the success rate all wells that are productive of some oil or gas are included. A few of the discoveries may later prove to be large hydrocarbon accumulations, while some may contain negligible quantities of oil or gas and others may prove to be economically non-commercial propositions.[1]

The gross success rate may therefore provide a misleading representation of the richness of the petroleum province. A more meaningful, though not error-free, guide is the average amount of oil discovered by wildcat drilling. Elf-Aquitaine, an oil company, has computed that in the decade 1960–70 the average quantity of oil brought in per exploratory hole in the free world excluding North America was 3.7 million metric tons, but only 1.68 million metric tons in countries outside the Middle East. Although not strictly comparable because the figures relate to different time periods, the quantity discovered per wildcat in Indonesia in 1974 was 1 million metric tons (Mainguy, 1976).

As soon as a field has been located by an exploration well, a number of step-out or appraisal wells are drilled to establish the extent of the field. Once it has been proved that a commercially viable accumulation exists in that field, development or production wells are drilled to allow a regulated and profitable flow of oil or gas to the surface. There is a time interval between the establishment of a commercial discovery and the drilling of production wells as permanent facilities have to be constructed on land or producing platforms built at sea. Figure 4.5 shows details of development wells drilled in Indonesia during the period 1966–79. The years 1966–70 saw little development activities, but from 1971 onwards a large number of wells were drilled each year, climaxing in 1975 when 327

[1]To be considered a 'commercial find' a new discovery must have a potential capacity of 300 to 1,000 barrels/day for onshore wells (depending on development costs) or 2,000 barrels/day for offshore wells (U.S. General Accounting Office, 1979, p. 7).

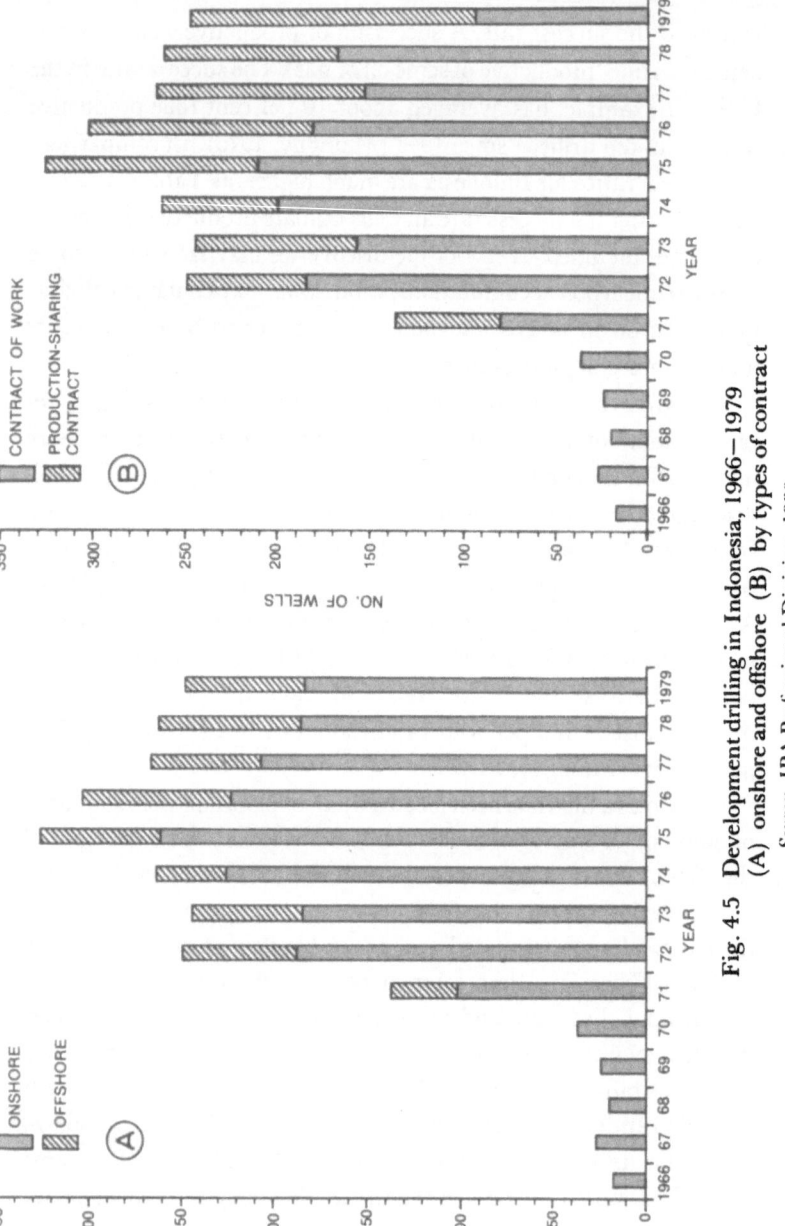

Fig. 4.5 Development drilling in Indonesia, 1966–1979
(A) onshore and offshore (B) by types of contract
Source: IPA Professional Division, 1980.

wells were drilled. Since 1971 nearly two development wells were drilled for every exploratory well drilled. Most of the development wells have been drilled in onshore areas by companies operating under Contracts of Work, but allowing for the time gap between discovery and development, the percentage of wells drilled in offshore areas by companies operating under production-sharing contracts has been increasing since 1975, and will increase with the development of fields in the Java Sea and the South China Sea.

The quantity of hydrocarbons that can be recovered in a field (the recoverable reserves) will depend on three factors: the physical characteristics of the deposit; technical factors; and prevailing economic conditions. The main physical characteristic influencing the recovery of oil or gas by primary recovery techniques is the type of natural-reservoir energy available to drive the oil out of the reservoir to the well and then to the surface. The most effective technique is water-drive, which depends on water under hydrostatic pressure forcing the oil or gas towards the well bore once a well has penetrated the reservoir. The quantity of oil-in-place that can be recovered by water-drive under ideal conditions can be as high as 85 per cent (Arps *et al.*, 1967). The other usual form of natural-reservoir energy is the gas that accumulates above the oil to form the gas cap and/or the solution gas in the oil. This is less effective than water-drive since the gas pressure will decrease as production proceeds. These primary recovery techniques, water-drive and gas-drive, were estimated by Clegg (1967) to provide average recovery rates of 27 per cent for oil. Improvements in primary recovery techniques have led to average recovery rates for oil increasing to 30 to 40 per cent in recent years. Recovery rates for gas are higher—from 60 to 80 per cent (Mainguy, 1976).

This implies that without additional effort, from 60 to 70 per cent of the oil in place and 20 to 40 per cent of the gas in place will remain unrecovered. In order to increase the recovery rates, secondary recovery techniques may be employed. These operate by artificially supplying energy to the reservoir, and act directly on the pay horizon. They consist mainly of water or gas injection. Pumping water into the reservoir beneath the oil will create a type of artificial water-drive which will force the oil towards the well. Similarly the gas-drive in a reservoir can be boosted by pumping gas into the reservoir to increase its gas pressure.

The need to install secondary recovery techniques depends on reservoir characteristics as well as economic considerations. When a reservoir is entirely surrounded by impervious rock the only natural-reservoir energy source comes from the decompression of the oil itself. The recovery rate is consequently very low. In such a situation a secondary recovery method is necessary.

But in most cases economic considerations govern the use of secondary recovery techniques in an oil or gas field. Since these techniques require additional investments their viability will hinge upon the magnitude of these investments, the quantity of additional oil that can be recovered and the returns that can be obtained from selling the oil. In past years secondary recovery measures were initiated when a field was almost at the economic limits of its primary production. But the practice today is to initiate these measures at an early stage in production in order to add to the total oil recovery.

In recent years advances in oil recovery technology as well as the steep rises in crude oil prices have made the use of enhanced recovery methods increasingly attractive to the industry. Enhanced recovery (or ER in oil jargon) is defined as oil recovery by processes aimed at higher displacement efficiencies than can be obtained through the natural processes of gas and water-drive. ER, in short, aims to increase the flow of oil to the producing wells and thereby raise the recovery rate.

ER methods include the application of heat to reduce viscosity and make the oil flow more easily (hot water injection; steam-soak; steam-drive; *in situ* combustion); the injection of gases or liquids such as carbon dioxide which act as a solvent for the oil and thereby increasing its flow; or the use of chemicals (polymers, miscellar-polymers, surfactants, caustic soda) to reduce interfacial tension, and improve oil flow.

The application of these methods is limited by a number of constraints. There is first the necessity to obtain detailed technical information about a particular oil reservoir, which information can only be provided by highly trained staff. Such staff are in short supply. Second is the risk of encountering some unexpected geological condition in the reservoir which could reduce the projected oil flow. Third, and most important, is the high level of capital investment needed for these sophisticated methods (Tucker, 1980). In 1979 the U.S. Department of Energy estimated the minimum prices of crude oil required to make the use of ER methods economically feasible to be as follows (*OPEC Bulletin*, 9 July 1979):

ER Method	Minimum Prices per Barrel (1979)
Steam-drive	$11 to $16
In situ combustion	$13 to $20
Carbon dioxide flooding	$13 to $23.50
Surfactant/Polymer flooding	$20 to $32

The rise in oil prices to above the $30 per barrel level has made the use of all but the most expensive chemical methods of ER an apparently economic proposition. On the grounds that it would be mutually beneficial to both the country owning the oil resources and the oil company extracting them to ensure that the recovery rates are maximized, the government could provide various fiscal or tax incentives to the companies to invest in ER schemes.

Such incentives are now provided by the Indonesian government in various forms. In the case of the Minas field operated by Caltex, the largest producer of oil in Indonesia, the government has agreed to give the company a reduction in its tax-paid cost for its share of the incremental oil. Incremental oil is the output above an agreed primary decline curve for the Minas field. The definition of this curve will be reviewed every two years (*Petroleum Economist*, May 1978). In the case of two other companies—Tesoro in East Kalimantan and Redco in Sumatra—working in areas previously producing oil, they will receive a 65/35 split on the incremental oil output resulting from the use of ER methods or from new exploration (American Embassy, 1978, p. 14). Total Indonesie is undertaking a $75 million ER programme in its Handil field, while IIAPCO has plans for a similar programme for its Rama field in the Java Sea.

Exploration Targets

The dramatic increase in oil exploration activities since 1966 has resulted in a better understanding of the geologic framework of hydrocarbon accumulation and this in turn has made for a more precise focusing of exploration efforts. New insights on the mechanism of basin formation, basinal structural framework and the environmental and depositional conditions of the Tertiary basins of Indonesia have established the basic reference for future prospecting in these basins and contributed to the conceptualization of the hydrocarbon-producing situations in the country.

One such concept, developed in 1975, is that of 'play-types' (Soeparjadi *et al.*, 1975). The play concept is defined as a combination of geological factors that result in a hydrocarbon accumulation. The four elements—reservoir, seal, source and trap—which are necessary to a play can combine in different ways to make up a diversity of play-types. The play-types developed by Soeparjadi and other petroleum geologists working in Indonesia are conceived of in terms of reservoir rock, with the other elements that are essential for hydrocarbon accumulation being present.

Based on previous exploration efforts and findings six play-types have been identified in Indonesia:
1. Transgressive Clastics
2. Regressive Clastics
3. Deltaics
4. Carbonate Platform Complex
5. Pinnacle Reefs
6. Fractured Volcanics/Basement

Figure 4.6 is a diagrammatic representation of the productive Tertiary plays in Indonesia, with a generalized vertical time-scale. Table 4.2 shows reserves, production and field size distribution of these play-types.

Transgressive clastics are sandstone reservoirs deposited during the major transgression of Eocene to Early Miocene time, and again in the Pleistocene when the sea-level rose. Source rocks are older or contemporary shales. These sediments were structured into large anticlines during the Plio-Pleistocene. These clastic reservoirs are

91

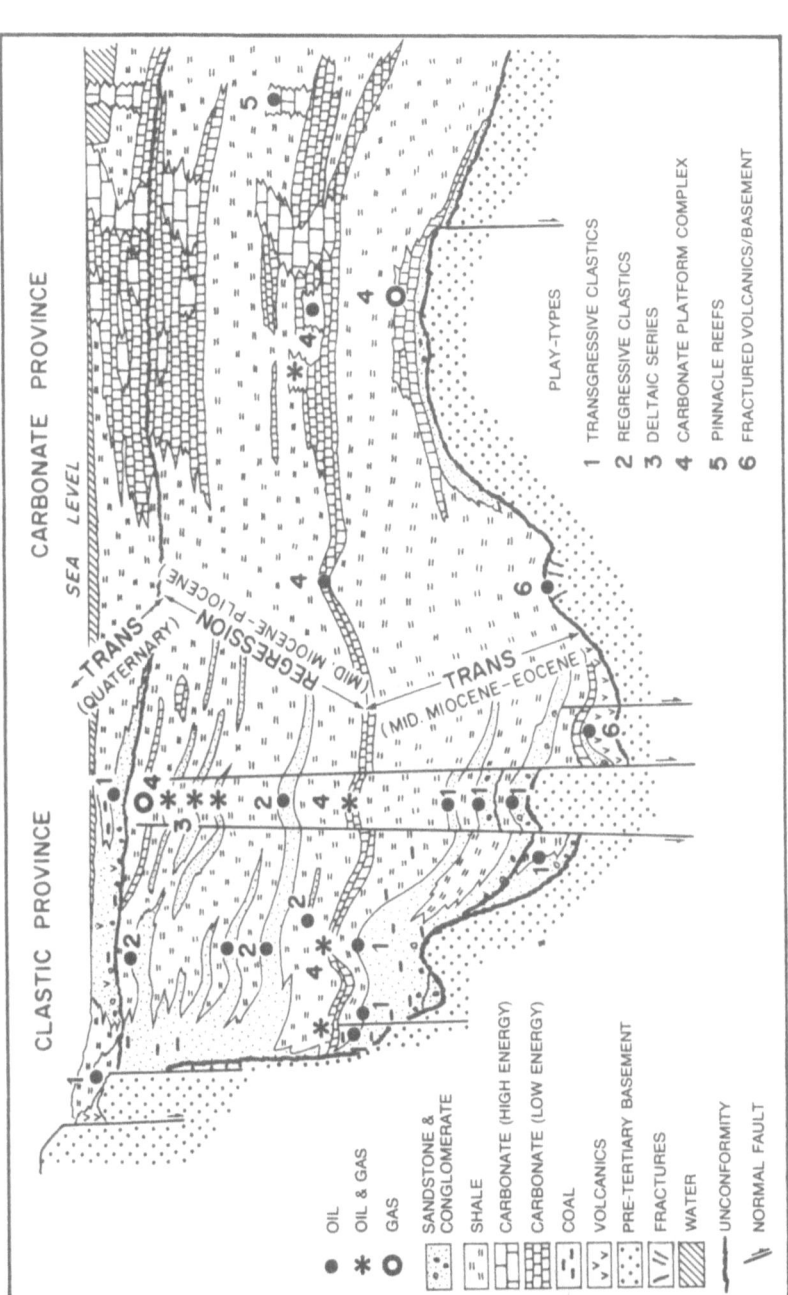

Fig. 4.6 Diagrammatic cross-section of producing Tertiary plays in Indonesia

Source: Soeparjadi, *et al.*, 1975.

TABLE 4.2
RESERVES, PRODUCTION AND FIELD SIZE DISTRIBUTION OF PLAY-TYPES IN INDONESIA

	Play-Type 1	Play-Type 2	Play-Type 3	Play-Type 4	Play-Type 5	Play-Type 6
Reservoir rock	Transgressive Clastics	Regressive Clastics	Deltaics	Carbonate Platform Complex	Pinnacle Reefs	Fractured Volcanics/ Basement
Reservoir age	Eocene to Early Miocene and Pleistocene	Middle Miocene to Upper Pliocene	Late Miocene to Upper Pliocene	Upper Oligocene to Late Miocene	Middle to Late Miocene	Pre-Tertiary and Oligocene
Estimated per cent of discovered oil reserves	62	11	10	14	2	1
Cumulative oil production up to 1975 (million barrels)	4,040	1,359 +	325 +	50 +	4 +	7 +
Number of basins with play-type	8	10	1	9	1	3
Range of field size:						
Super-giant (1,000+ million barrels)	6%	—	11%	9%	—	—
Giant (500—1,000 million barrels)	2%	—	—	—	—	—
Major (50—500 million barrels)	25%	29%	44%	—	67%	50%
Minor (1—50 million barrels)	67%	71%	45%	91%	33%	50%
Number of fields used in estimates of field size	54	38	9	11	6	2

Source: compiled from Soeparjadi *et al.* (1975), Tables 1 & 2.

highly prolific and have been responsible for the greater part of Indonesia's crude oil production. They are best represented in the Central and South Sumatra basins. More recently, oil has also been discovered in transgressive clastics in the Sunda basin, offshore South-East Sumatra.

Regressive clastics are finer-grained sands laid down during the periods when the seas retreated. These reservoir rocks were sourced by adjacent marine shales and are found mainly in structural traps formed in Plio-Pleistocene time. They are the second most important source of oil in Indonesia, and are found in the North (Aceh) and South Sumatra basins, the North-West and North-East Java basins and the East Kalimantan Tertiary basins.

Deltaics are locally important deposits of quartz-rich sandstones laid down during a regressive phase, but some Early Tertiary deltaic sediments in the South Sumatra, Sunda and North-West Java basins have also been grouped with the Transgressive Clastics (Soeparjadi *et al.*, 1975). The most important deltaic complex is that in the Kutei basin. Of Late Miocene to Recent age, capped and sourced by adjacent shales, these sandstones contain both oil and gas in structural and stratigraphic traps.

Carbonate platform complex comprises reefoid and bioclastic reservoirs developed upon a carbonate platform shelf, with source rocks within, or overlying or adjacent to the porosity zone. Each porous body is usually small but a number of these reservoirs may be located together to form a large accumulation. The giant Arun gas field in the North Sumatra (Aceh) basin is an example of such an accumulation. Carbonate reservoirs are now a prime target in the search for hydrocarbons in several basin areas.

Pinnacle reefs are isolated vertical features which have developed on the seaward edge of carbonate platforms during a period of active basin subsidence, or which have developed in local basinal deeps. Pinnacle reefs have proved to be productive in the Salawati basin of Irian Jaya, where Middle to Late Miocene reefs and associated porosity are found in stratigraphic traps of shale, and are sourced by adjacent marine shales. The Salawati pinnacle reefs have high production rates, with hydrocarbon columns of up to 130 m but with reservoir areas of only between 240 to 400 ha (Vincelette, 1973).

Buried pinnacle reefs are currently the exploration targets in the onshore and offshore areas of Irian Jaya and the Java Sea.

Fractured volcanics/basement reservoirs are, as the term indicates, fractured reservoirs of volcanic origin—Oligocene andesitic tuff and tuff breccias and pre-Tertiary volcanics and extrusives. They are found in folded structures sealed and sourced by overlying clastics. Two onshore fields, Jatibarang in the North-West Java basin and Tanjung in the Barito basin, East Kalimantan, belong to this play-type.

As stated earlier the hydrocarbon prospects for Indonesia lie mainly in the Tertiary. The status of exploration of the onshore and offshore Tertiary areas up to the early 1970s is indicated in Table 4.3 below.

More recently Nayoan and other Indonesian petroleum geologists have identified 40 Tertiary basins in Indonesia, of which 10 are classed as intensively explored, 11 moderately or partially explored, and 19 unexplored (Nayoan *et al.*, 1979; Hariadi, 1980). Figure 4.7 shows their distribution.

Exploration targets and hydrocarbon finds in Indonesia have been from Miocene sandstones in the post-war period. Until 1968 sandstone was the only reservoir source found in Indonesia. But subsequent discoveries of large hydrocarbon accumulations in Miocene carbonates—in Arun, North Sumatra; in the North-West and North-

TABLE 4.3

EXPLORATION IN TERTIARY AREAS, INDONESIA

	Onshore (sq. km)	Offshore (sq. km)	Total (sq. km)
Prospective area with Tertiary sedimentary cover	1 700 000	3 400 000	5 100 000
Extensively explored	172 000	—	172 000
Presently being extensively explored	260 000	1 900 000	2 160 000
Awaiting extensive exploration	1 500 000 (deeper areas)	1 268 000	2 768 000

Source: Akil and Nayoan, 1973.

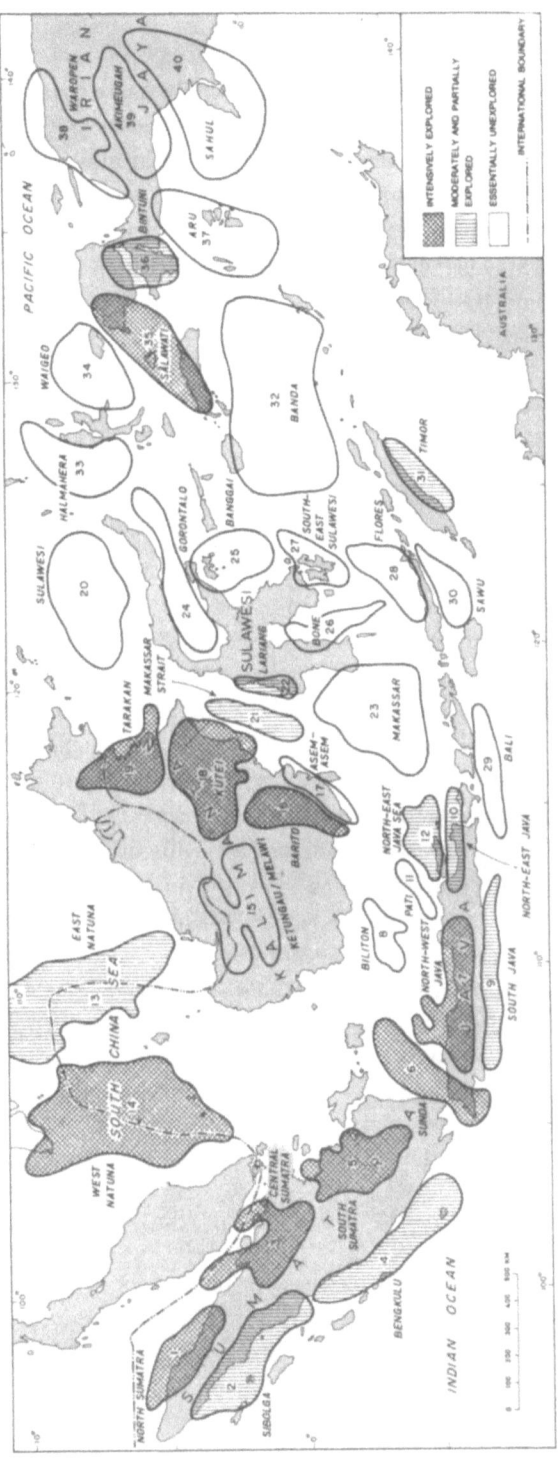

Fig. 4.7 Exploration status of the Tertiary basins in Indonesia
Source: Nayoan, et al., 1979; Hariadi, 1980.

East Java basins; in the Sunda basin; and in the Salawati basin—have drawn attention to the large potential from these rocks. Carbonates, mainly limestone, are believed to be present in most of the Indonesian islands, particularly in the eastern and central areas. These carbonate reservoirs (designated as play-types 4 and 5 by Soeparjadi *et al.*, 1975) are currently one of the main exploration targets in most of the Tertiary basins of Indonesia, particularly in the Java Sea and Irian Jaya where the search is focused on buried pinnacle reefs. In other basinal areas patch reef, bank edge and bioclastic porosity are the targets. In mid-1974 oil production from pinnacle and patch reef reservoirs in Irian Jaya and offshore North-West Java averaged 50,000 barrels per day (Soeparjadi *et al.*, 1975). By 1978, 20 per cent of the total Indonesian oil production was derived from carbonate reservoirs, as against 78 per cent from sandstone reservoirs (PERTAMINA, 1979).

A third reservoir source, first discovered in 1968 at Jatibarang, West Java, is from volcanic clastics (designated as play-type 6 by Soeparjadi *et al.*). Other discoveries have since been made. Although crude oil from volcanics constituted only 2 per cent of 1978 production, Indonesian geologists believe they have an important potential in view of the wide distribution of such rocks in Central and West Indonesia. The target here is for fracture porosity in volcanics as well as possible intergranular porosity in tuffs and breccias (Soeparjadi *et al.*, 1975; PERTAMINA, 1979).

Of the clastic reservoirs the sand—shale sequences identified as deltaic reservoirs have been prolific sources of oil and gas in the Kutei basin of Kalimantan, and these as well as transgressive and regressive clastics are prime exploration targets today.

Attention is also being directed to deep water areas on the continental slope and rise which may contain carbonate as well as other reservoir rocks. Hariadi (1980) has identified fifteen Tertiary basins in Indonesia which lie in waters deeper than 200 m and which are unexplored. As the technical and economic limitations to drilling and production recede with technological advancement and oil price escalations, the deep water areas will begin to attract increasing interest as a potential source of hydrocarbon accumulations.

Exploration Phases

It is possible to divide the history of oil exploration and production in Indonesia into fairly distinct phases. As has been shown earlier, the history of the Indonesian oil industry is a long one, going back as far as the 1880s. Exploration in the first pioneering phase covering the years from the 1880s to the early 1930s was, by necessity, confined to onshore locations, mainly in the larger islands of Indonesia. During this phase, all concessions up to 1924 were granted for a period of seventy-five years. Within fifteen years after the first discovery was made in North Sumatra, the basins in South Sumatra, East Java, East Kalimantan and Tarakan island in North-East Kalimantan were producing oil. In 1911 the area held under oil concessions amounted to 3 200 sq. km; by 1924 this had increased to 6 400 sq. km. Most of the producing areas found during this first phase are today still producing small quantities of oil, amounting in all to an estimated 3 per cent of total production in the mid-1970s (Fletcher & Soeparjadi, 1976).

The second phase may be described as the modern onshore exploration phase, extending over the thirty-five years from 1930 to 1965. Two major developments occurred during this period. The first involved a change in the conditions under which concessions were granted. The new contracts were for forty instead of seventy-five years. The contracting company had an obligation to drill but could return to the state those parts of its area that proved barren. The state now would receive royalties for the concessions as well as a progressive profit share. The second development was the technological progress achieved in exploration techniques, of which the most important was the use of aircraft for photographic surveys from 1934 onwards. Such surveys could cover large expanses of difficult swamp and jungle-clad country at the fraction of the time needed for ground surveys, and at a much lower cost. Between 1934 and 1940 the oil companies mapped an area of 175 000 sq. km in Sumatra, Kalimantan and Irian Jaya. Up to the mid-1930s exploratory work was confined to the main islands of Sumatra, Java and Kalimantan but, as the pace of exploration increased in Indonesia, the search for oil was extended to Irian Jaya, and some 100 000 sq. km of this remote territory were surveyed just before the outbreak of the Pacific War (Pratt & Good, 1950).

The post-war years were directed towards the rehabilitation of production facilities destroyed or damaged during the war, followed by active exploration onshore in the Aceh basin in North Sumatra, the South Sumatra basin, the East Java basin, the Barito and Kutei basins of Kalimantan, and in the island of Seram. The most notable discovery of this period was the giant Minas field in Central Sumatra. Towards the end of this modern onshore phase there was a reduction in the oil companies investments in exploration activities because of the political uncertainties following the Indonesian Communist Party's abortive coup of 1965 as well as changes in the relationships between the Indonesian government and the foreign oil companies. In addition, external factors such as the erosion of oil prices in the early 1960s due to surplus world oil production and the concentration of the oil industry's funds in major exploration efforts elsewhere in the world acted as depressants to exploration activity in Indonesia. Such exploration work as was conducted after the war was confined to onshore basins in Sumatra, Java and Kalimantan (Fig. 4.8). Discoveries made during these thirty-five years of modern onshore exploration contributed about 47 per cent of Indonesia's total oil production of 1,370,000 barrels per day in 1974. The most significant discoveries of this period were the two fields, Minas and Duri, which together were responsible for almost 30 per cent of Indonesia's production in 1974.

The second modern exploration phase may be said to start from 1966 when Indonesia's first offshore production-sharing contract was signed between IIAPCO and the Indonesian government. During this period the number of operating oil companies in Indonesia increased from eight in 1966 to thirty in 1975. This large increase was a reflection of the oil companies' interest in exploring the offshore basins of Indonesia, now made possible because of advances in offshore exploration and drilling technology and equipment. The rapid acceleration in exploration activities may be seen in the more than tenfold increase in the sums expended on oil and gas exploration and development in Indonesia between 1969 and 1974—from $78 million to $807 million.

Fig. 4.8 Exploration foci in Indonesia, 1946–1966
Source: After Hatley, 1976.

Exploration work was mainly geophysical in nature, mostly seismic reflection surveys combined with some geological field-work. Some surveys were preceded by aerial surveys. The rapid acceleration in exploration activities may also be seen in the extent of surveys carried out: in the years 1970 to 1972 an annual average of 80,000 line km of land seismic, shallow and deep water seismic surveys were conducted in Indonesia. This was about twenty times the annual average of the pre-1966 period.

Most of the exploration work during this phase was carried out in offshore areas. The concentration of attention on offshore areas is seen in the percentage figures of offshore exploration activities in relation to total exploration activities during the years 1972 to 1974: 84 per cent in 1972, 78 per cent in 1973 and 79 per cent in 1974. However, extensive onshore operations were also carried out, many of them in remote areas made accessible through the use of helicopters. The distribution of exploration activities during this phase is indicated in Figure 4.9. Significant discoveries were made in both onshore and offshore locations in Sumatra, Java, Kalimantan and Irian Jaya (Fig. 4.10). The sum total of these efforts was an additional production (in 1974) of 300,000 barrels per day from offshore discoveries in the Java Sea and East Kalimantan and 400,000 barrels per day from onshore discoveries, mainly in Sumatra, as well as several billion barrels of reserves. Two very large gas fields were also discovered, at Arun in the northern coastal plain of Sumatra, and at Badak in the Balik-papan area of Kalimantan.

This phase of accelerated exploration activity (termed the 'quiet boom' in oil circles) came to an end in 1974 when there was a sharp downtrend in oil exploration, not only in Indonesia but throughout the world (Hatley, 1976). The downtrend in Indonesia is illustrated in Figure 4.4 which shows the reduction in the number of exploratory wells drilled between 1975 and 1977. The geographical focus of exploration attention shrank correspondingly from the very large, mainly offshore, area shown in Figure 4.9 to those sections of Java, East Kalimantan and western Irian Jaya delimited in Figure 4.11.

A complexity of factors, external as well as internal, have contributed to the decline in exploration activity in Indonesia. Hatley (1976) has analysed in some detail the many external factors that

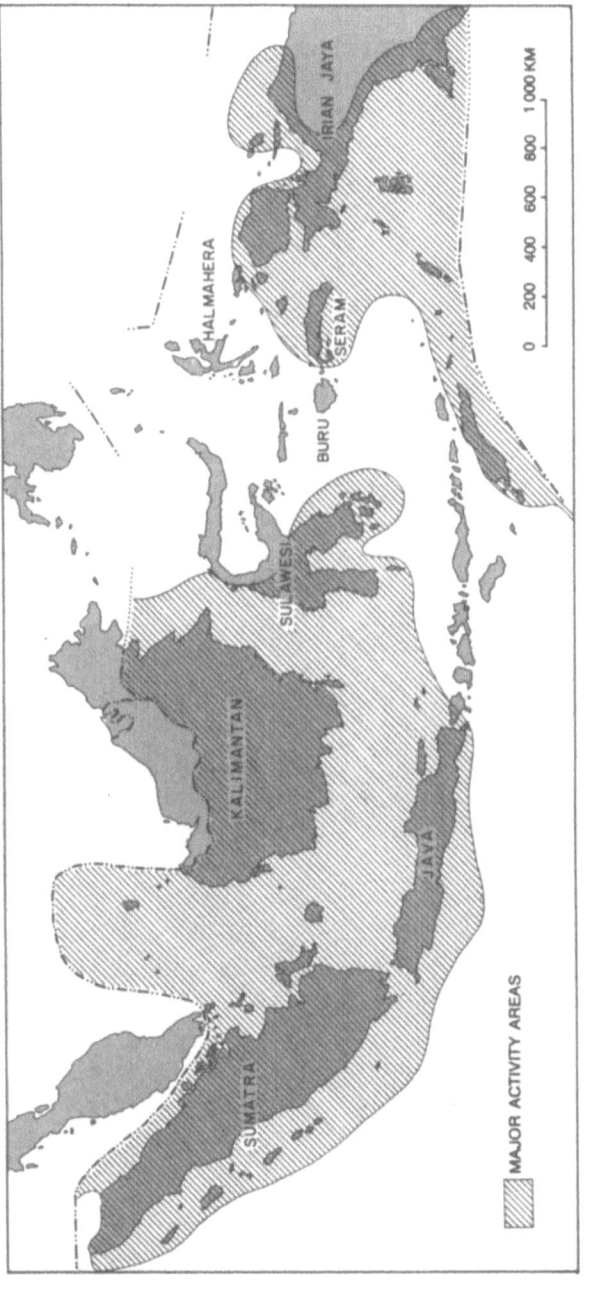

Fig. 4.9 Exploration foci in Indonesia, 1967–1974
Source: After Hatley, 1976.

Fig. 4.10 Oil fields discovered in Indonesia before and after 1966
Source: Soediono and Loucks, 1976.

Fig. 4.11 Exploration foci in Indonesia, post-1974
Source: After Hatley, 1976.

have affected the world oil industry. Among these were the basic changes that occurred in the structure of the oil industry in the early 1970s when the privately-owned oil companies lost control of the world's major reserves and producing capacity of petroleum to the Organization of Petroleum Exporting Countries (OPEC). This in turn paved the way for the 1973−4 oil embargoes and the subsequent oil price increases. The result was a worldwide reduction in oil consumption and consequently a decline in oil sales.

Furthermore the oil price increases served to deepen the economic recession among the industrial countries, in particular the U.S., Japan and some European nations, which countries are also the main sources of investment capital, advanced exploration and production technology and expertise in the oil industry. This fact, coupled with the sharply reduced profitability of investing in the oil industry through various host governments' measures to increase their take on each barrel of oil produced, seriously affected the international oil companies' capacity and willingness to enter into new exploration ventures. Such ventures as were undertaken were on a highly selective basis, with exploration money going into low risk prospects offering the most attractive potential rates of return, such as those in Alaska and the North Sea.

These factors have also had repercussions on exploration activity in Indonesia, since it is a country heavily dependent on international, mainly American oil companies for investment and development capital. Such companies must take into consideration the climate for investment at home as well as in those overseas countries in which they operate. In this connection U.S. companies operating in Indonesia were confronted with a major impediment to further exploration and development when the U.S. Internal Revenue Service (IRS) issued a ruling in May 1976 disallowing a foreign tax credit for a share of oil production retained by the Indonesian government under a production-sharing contract with a U.S. oil company. The ruling did not affect companies which entered into production-sharing contracts before 8 April 1976. The basis of the ruling was that a tax cannot be considered to have been paid by a company to Indonesia from its (that is, Indonesia's) share of production because such share of production was always the property of Indonesia and was not acquired from the company. The Indonesian share of oil from production-sharing agreements was therefore a royalty payment, and as such U.S. oil companies could qualify for only a 50 per cent reduction in tax and not a 100 per cent reduction. The effect was to reduce the oil companies' share from 15 per cent (based on the 85/15 split) to between 7 and 8 per cent. The problem was resolved in May 1978 when the IRS decided that payments made by U.S. oil companies to the Government of Indonesia were legitimate income tax payments and were therefore eligible for foreign tax credit in the U.S.

The PERTAMINA financial crisis of 1975 and the measures taken by the government to solve it has been discussed in Chapter 2. As seen earlier, one of the measures was to revise the contractual terms under which the foreign oil companies operated in Indonesia so as to increase the governmental share of the oil proceeds. The immediate consequence of this move was for the companies to call a halt to exploration activities while they reassessed the investment climate in Indonesia.

Gaffney *et al.* (1976) have carried out a study that showed that a split of 85/15 could have the effect of reducing the level of reserves discovered in South-East Asia by about one-third over the next decade, while an average split of 95/5 at current oil prices would bring further exploration by the oil companies to a complete stop.

Low exploration activity will affect Indonesia's capacity to maintain production levels. Indonesia's oil reservoirs are small by Middle East standards and exploration is essential to bring new fields into production as the flows from existing fields taper off and eventually dry up.

However there are a number of factors that have to be borne in mind when discussing the level of exploration activity. First is the fact that Indonesia's petroleum resources have a crucial role to play in the development of its national economy. Oil not only dominates the mining sector but it constitutes the most valuable export commodity and is the main source of foreign exchange earnings. In order to generate the capital for its development efforts Indonesia has generally followed a policy of extracting its oil out of the ground as quickly as possible. To do this it has had, and still has, to rely largely on foreign capital and expertise in the oil sector. In seeking to maximize its oil earnings Indonesia has nevertheless to pay attention to the need to provide a reasonable reward to the foreign oil companies for the risks they bear in their search for oil. A poor investment climate may well drive the oil companies to seek pastures elsewhere. Too liberal conditions would reduce governmental oil earnings to a less than maximum level. The balance between the two is often a fine one.

It is apparent that the governmental revision of contracts with the oil companies in 1976 had tipped the scale against the companies. As the former Mines Minister Mohammad Sadli said in a speech at the Sixth Indonesian Petroleum Conference on 23 May 1977:

allow me to make a few comments on the renegotiation of the terms of the oil contracts. First of all we do realize that the action has shocked the offshore oil industry, so much so that exploration activity last year has decreased considerably. To the extent that this was unavoidable because of reduced cash flows in the short run, we accept this as a fact of life. To the extent that the terms are falling short of the requirements for new exploration we try to remedy the situation. The reasons and motives of the Government decisions are now well known. The main reason was one of economic justice. The large economic rent that should accrue to host governments as a result of the substantial increases in oil prices should be properly reflected in the new terms of the contract. The second reason was that the country was in a balance of payments and financial predicament in the wake of Pertamina's debts. The change of leadership in Pertamina also affected the decisions. In the process of renegotiations the Government and Pertamina learned a lot about the character of the Indonesia oil industry, about its requirements for exploration and development. This learning process has resulted in a more flexible attitude of the Government and Pertamina. It is imperative that the exploration and development climate be preserved. It is true that oil contract policies are now more complex and refined than before; this is to ensure optimal government revenues on the one hand and the maintenance of an effective exploration and development climate on the other hand. Concepts like old oil and new oil are introduced, fiscal formulas for giving greater front end relief for the companies are applied. The degree of difficulty to find and lift the oil is, or will be, reflected in the formulas as the Government and Pertamina gain more insight through frank exchanges and dialogue.

The new terms of the renegotiated contracts, which were made retroactive to 1 January 1976, produced an additional $550 million in governmental revenues. But as the oil companies' profit margin was reduced substantially, they have had to review their current and future exploration and development programmes. Many, especially those with production-sharing contracts, ceased their search for oil and the number of exploration wells fell quite significantly (see Fig. 4.4). The fall was mainly in the offshore sector, the number of offshore rigs operating in Indonesian waters decreasing from 11 in July 1976 to only 4 in September 1976 (Direktorat Jendral, 1978).

Reduced exploration and drilling activities by the foreign oil companies put PERTAMINA and the government in a quandary as the national oil company had no offshore drilling capacity and was dependent on the foreign oil companies to sustain and expand production.

Some industry sources have maintained that the contract revisions could not be held entirely responsible for the diminished level of oil exploration in Indonesia as the trend was already discernible before the contract revisions. The reasons cited for the downtrend were the reduced amount of risk funds available to the industry for exploration, heavy commitments to Alaskan and North Sea developments by some major oil companies, and changes in U.S. exploration incentives. Nevertheless the contract revisions had served to enhance the trend, especially in the case of the companies that were heavily dependent on their Indonesian operations, and those dependent on internal generation of exploration funds.

In an attempt to reduce this problem the government in early 1977 announced new incentives designed to stimulate oil exploration. These included a reduction of the government tax rate on future oil discoveries and measures to improve depreciation allowances. Three main concessions were involved. First, modifications were made to the original pro rata obligation of the oil companies to provide a portion of their production to PERTAMINA for Indonesian domestic consumption on the basis of extraction cost plus 20¢. The main change was that all pro rata domestic consumption oil provided by the oil companies would now be valued at the prevailing market price for a period of five years, the extra payment to be used for continued exploration in the companies' contract areas. If however, a company could demonstrate that no such opportunity for exploration exists, it would be allowed to use the funds as it saw fit (American Embassy, 1977).

Second, the government would also provide a 20 per cent credit for capital investments for the development of new oil fields in areas where start-up costs are higher than average. These areas are the deep water areas (water depth of over 300 ft) or onshore areas in remote locations (more than 30 mls from the coast or the nearest oil terminal or pipeline). This investment credit may be taken up by the oil companies in the first year of new production, before they release 85 per cent of their 'profit oil' to the government under the terms of their production-sharing contracts. This investment credit incentive was later extended to all companies undertaking new exploration, provided the government receives at least 49 per cent of the cumulative equity in production over the life of the particular field.

Third, production-sharing contractors would be permitted to recover their capital investment over a period of seven years using a double declining balance depreciation system, on all investment schemes related to the development of 'new' oil.

In the case of companies operating under Contracts of Work, the government announced that, from 1 January 1977, the companies could withhold 50¢ per barrel for oil produced from new fields or from existing, depleting fields where secondary recovery methods are employed. In May 1977 PERTAMINA and Caltex signed a five-year agreement incorporating these incentives.

The foreign oil companies have responded to the exploration incentives extended by the government as well as to the high oil price levels by increasing their investments in oil exploration from $140 million in 1977 to $250 million in 1978 and to an expected $380 million in 1979 (American Embassy, 1980, p. 20). The upturn in exploration was also reflected in the increase in the number of exploration wells in the period 1978 to 1979 (Fig. 4.4).

The mid-1970s phase of low exploration activity in Indonesia was thus not more than a temporary shift in position. The symbiotic relationship between the capital and technological skill of the foreign oil companies, on the one hand, and the economic need of Indonesia, on the other, has operated to move exploration activity out of the doldrums as both the companies and the government adjusted themselves to the changed external and internal conditions that had caused a drop in exploration levels.

In anticipation of the projected demand for new exploration areas, PERTAMINA invited foreign oil companies to search for oil in onshore and offshore areas previously reserved for itself. Operators in these areas will have to sign a new form of contract—the joint venture or '50-50' contract—whereby they must match PERTAMINA's pervious expenditures in the contracted area or bear all costs for the first three years of exploration, whichever is greater. Beyond the amount, PERTAMINA and the contractor will share all costs equally and divide up any production also equally. But the contractor's share of production is then split 85/15 in favour of PERTAMINA.

Three of the four blocks released in 1978 by PERTAMINA for joint-venture contracts had been taken up by the end of the year: Deminex, a West German oil company, contracting for the 4 285 sq. km Simenggaris onshore block in North-East Kalimantan; JAPEX for the 21 000 sq. km Lampung onshore block in South Sumatra; and Teikoku Oil for the 5 335 sq. km Jambi A block onshore in south Central Sumatra. The 3 400 sq. km Pensangan block offshore North Sumatra was awarded to Mobil in 1979. Three other blocks were contracted out to oil companies in 1979. Figure 4.12 shows the oil and gas contract areas, onshore and offshore, as at January 1980. Details of the contract areas are summarized in Appendix A.

In mid-1979 Indonesia and the Japan National Oil Corporation (JNOC) signed a cash-for-oil agreement whereby the JNOC will provide Indonesia with a $310 million loan for oil exploration which would be repaid not in cash but in oil. The funds would be used to finance exploration in four onshore areas in Java, Kalimantan, North Sumatra and South Sumatra. Repayment will only be made if oil is discovered and developed; the funds would be forefeited if no oil is found.

If oil is discovered and developed Indonesia will repay both the principal and a yearly interest charge of 6.5 per cent on the loan, calculated on the basis of prevailing world market prices. Repayment will be in oil equivalent. Indonesia would also allow the JNOC to buy up to 50 per cent of the production from these areas at world market

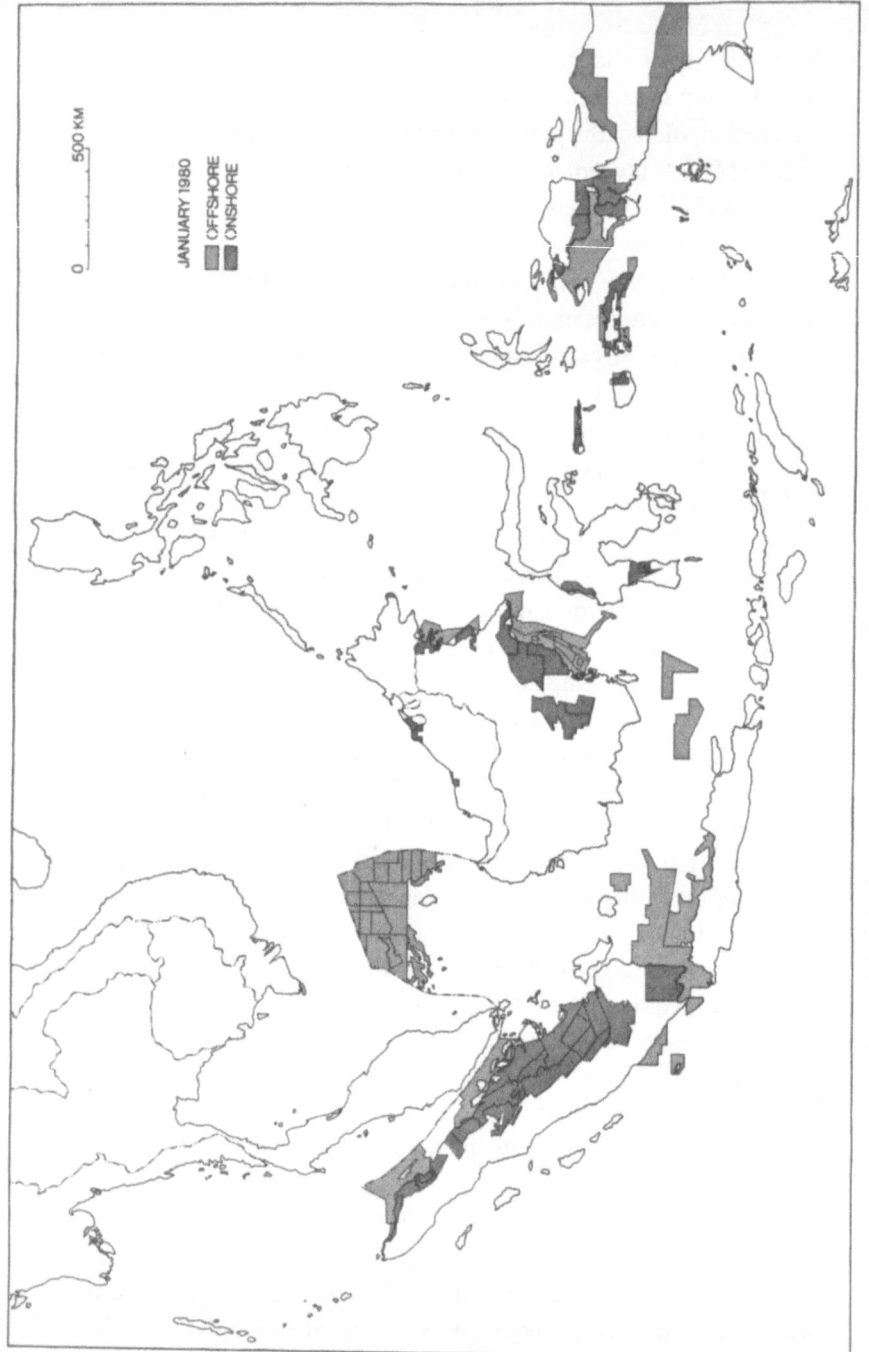

Fig. 4.12 Oil and gas contract areas in Indonesia, January 1980

prices. The JNOC would also receive 4.7 per cent of the production free of charge in areas it helps to explore and develop and 2.5 per cent of the production in areas where its funds are used to develop reserves already established through previous PERTAMINA exploration work (*Far Eastern Economic Review*, 29 June 1979; *Asia Research Bulletin*, 31 August 1979).

This agreement would provide PERTAMINA with immediate benefits in providing it with funds for continued exploration work, at no financial risk to itself. Japan, on the other hand, stands to gain only if oil is discovered in commercial quantities in the designated areas, through securing a source of oil supplies in a politically stable country which is also geographically closer to it than the Middle East.

The period 1979 to 1980 has not witnessed any revisions in the terms of the production-sharing contracts. However the most recent 1980 contracts contain a clause requiring that in the event of commercial oil production the foreign contractor offers 5 to 10 per cent of his interest to a 'national entrepreneur of Indonesia' (American Embassy, 1980, p. 20).

The search for oil in Indonesia reached its nadir in 1977 at the same time as production peaked. Since then the level of exploration activities has increased substantially, as seen in the 170 per cent increase in the exploration budgets of the foreign oil companies between 1977 and 1979.

Several factors have contributed to this uptrend in oil exploration. First, the various incentives offered by the government have proved to be sufficiently attractive to the oil companies for them to invest capital in low and middle risk exploration, in particular in exploration in the contract areas which have an existing production base.

Second, the political stability of Indonesia has made it an increasingly attractive alternative to the Middle East in the foreign oil companies' search for oil, more so after the recent upheavals in Iran and the uncertainties caused by the Iran-Iraq war.

Third, large parts of Indonesia are still little explored, with only eight of the twenty-eight Tertiary basins producing oil, and only the Central Sumatra basin being exploited intensively.

Fourth, the progressive increases in oil prices, to over $30 a barrel in 1980, have served as a stimulus to exploration and development, and have made previously marginal fields into economically viable propositions.

Fifth, the May 1978 announcement by the U.S. Internal Revenue Service that payments made by U.S. oil companies to the Government of Indonesia under the terms of production-sharing contracts are legitimate income tax payments and are eligible for foreign tax credit in U.S. has removed the oil companies' apprehension that they would be subject to double taxation in their Indonesian activities. This has in turn paved the way for renewed exploration and development spending by the U.S. oil companies which form the majority of the oil companies operating in Indonesia.

Sixth, the political leadership and petroleum officials of Indonesia have sought to engender a climate of confidence in Indonesia by repeatedly assuring the foreign oil companies that their investment in petroleum exploration and development would be protected. As President Suharto stated on 7 April 1979: 'I wish to stress that Indonesia still welcomes cooperation with foreign companies on a mutual basis. I wish to particularly stress to foreign companies that this cooperation will entitle them to a sound profit' (American Embassy, 1979, p. 28).

The search for oil in Indonesia has so far been confined to onshore and continental shelf areas. As would be expected the focus of attention in the continental shelf has been, up to now, on prospective areas located in relatively shallow waters of between 100 and 150 m. Development costs (drilling and production installations), which increase sharply with water depth, are not prohibitively high in these shelf areas, though they are higher than in onshore locations.

The history of offshore oil development in Indonesia goes back to only just over ten years, and many of the prospective shelf areas in relatively shallow waters are yet to be fully explored. As oil prices, which in early 1980 were over $30 per barrel, continue their upward climb the search for oil will cover not only the unexplored basinal areas of the continental shelf but will in time include areas which were once considered marginal.

Among these new zones of interest are the sedimentary basins of Indonesia located in the deeper waters of the continental slope and rise (see Fig. 3.5). Some of these basins are known to contain Tertiary sediments more than 6 000 m thick, but most have yet to be drilled or even properly mapped. Basins so identified are those offshore of West Sumatra, off South Java, off the eastern part of Sulawesi, in the Banda Sea area and off the northern coast of Irian Jaya (*Pertamina Bulletin*, July 1979). Although technological progress in exploration and drilling in deeper waters have been and will doubtless continue to be achieved, costs remain high (see Leuch and Masseron, 1973; Mainguy, 1976) and the deeper offshore basins are likely to continue to be zones of high risk, requiring heavy capital investment and sophisticated equipment. Nevertheless the oil companies had shown interest in these areas as early as 1972–3 when Shell Oil drilled two wildcat wells in 350 and 364 m of water south of Java, and Mobil Oil was awarded the first deep-water production-sharing contract, off the Kutei basin of Kalimantan, where the water is over 200 m deep.

Physically and in other ways Indonesia remains an attractive country to the oil companies. The prospective area, both onshore and offshore, is very extensive, while the environmental conditions for offshore prospecting and development are mild compared with say, the North Sea. Furthermore, the record of success in Indonesia is high by world standards. It also possesses the four basic prerequisites considered by Hatley (1974) to be essential for a sound and rational exploration and development programme, namely, (1) a realistic government policy supported by effective petroleum legislation that will promote exploration and development; (2) available risk capital, normally from the private sector, to invest in these programmes; (3) a reasonable expectation of success for all participating parties; and (4) technology and good exploration judgement of those responsible for conducting the programmes.

Indonesia is now (1980) on the threshold of a major upswing in exploration activity that could, give its traditional high success ratio, result in arresting the declines in production in the last two years. The number of production-sharing contracts is forecast to increase to fifty in 1981, and the exploration commitments contained in these contracts as well as exploration work in existing producing contract areas should maintain the momentum of the search for oil for several years ahead. PERTAMINA is optimistic that the total exploration expenditures in 1979–80 to 1983–4 will rise to the $2 billion level it anticipates will be needed to boost crude oil production to the 1.8 million barrels per day in 1983–4 projected under its Third Five-Year plan (*Financial Times*, London, 12 June 1979).

Offshore Boundary Problems

Indonesia has common offshore boundaries with eight countries: Malaysia, Singapore, Thailand, the Philippines, Vietnam, Papua New Guinea, Australia and India. As the oil potential of its continental shelf became evident and oil companies' interest in offshore areas increased, it was necessary for the government to define and delimit its offshore boundaries in order that exploration could proceed without hindrance. The issues that bear on this are the archipelagic principle to which Indonesia subscribes, the territorial sea, and the continental shelf concept.

Indonesia, in its Government Regulation in lieu of law, dated 18 February 1960, claims that its archipelagic waters, as defined by a number of straight lines connecting the outermost points of the outermost islands of its archipelago, are its internal waters. These straight baselines totalled 8,167.6 nautical miles and enclose 666,000 sq. nautical miles of internal waters (Prescott, 1975, p. 106). The Indonesian territorial sea is a maritime belt of 12 nautical miles as measured from these baselines. The development of mineral resources in the archipelagic waters as internal waters as well as the 12-mile territorial sea is the exclusive preserve of Indonesia.

Law No. 11 of 1967 on the Basic Provisions of Mining defines Indonesian mining jurisdiction as 'the entire Indonesian archipelago, the land below the Indonesian waters and the continental shelf of the Indonesian archipelago'. The position of Indonesia towards the development of the continental shelf resources is clearly indicated by the proclamation made by President Suharto on 17 February 1969, as follows (ECAFE Doc. I&NR/PL/C.R. Paper 24):

The Government of the Republic of Indonesia,
Considering:

. . .

c. that it has become customary practice of States and justified by International Law that a coastal State has exclusive sovereignty and juristion over the mineral and other resources in the seabed and subsoil of the continental shelf, or its analogue in an archipelago, adjacent to, but outside, its territorial sea to a depth of 200 metres or, beyond that limit, to where the depths of the superjacent waters admit of the exploitation of such resources;

d. that, based on the above-mentioned considerations, it is necessary to announce a governmental Proclamation on the Indonesian Continental Shelf; Proclaims that:

1. All mineral and other natural resources, including living organisms of sedentary species, found in the seabed and subsoil of the continental shelf outside the area of Indonesian waters as defined in the Law No. 4 of 1960, down to a limit where the superjacent waters admits its exploitation and undertaking, are the property of Indonesia to determine and are under its exclusive jurisdiction.

2. In cases where the Indonesian continental shelf, including the depressions on the continental shelf, of the Indonesian archipelago has a border with another State, the Government of the Republic of Indonesia is prepared through negotiations with the State concerned to determine a border line in conformity with legal and equitable principles.

3. Until the existence of such agreement, the Government of the Republic of Indonesia will issue permits for exploration as well as for production of oil and natural gas and for the exploitation of mineral and other natural resources only for the territory on the Indonesian side of the median line drawn between the coast of Indonesian outermost islands or, in the case where the Indonesian territory on an island is adjacent to another State, on the Indonesian side of a line the points of which are equidistant from the nearest points on the baselines of the territorial sea of the respective State.

This declaration was based on the Geneva Continental Shelf Convention of 1958, and was later enacted into Law No. 1, 1973 of Indonesia.

The Third U.N. Law of the Sea Conference has reached a general concensus that the outer limits of the continental shelf shall be established at a distance of 200 nautical miles measured from the territorial sea baselines (without regard to the geomorphological features of the shelf), or when the continental shelf extends beyond 200 nautical miles, the outer limits shall be the outer edge of the continental margin.

In delimiting its boundaries with its neighbouring states Indonesia takes the position as stated in Article 6 paragraph 1 of the 1958 Geneva Convention on the Continental Shelf: 'In the absence of agreement and unless another boundary line is justified by special circumstances, the boundary is the median line every point of which is equidistant from the nearest points of the baselines from which the breadth of the territorial sea of each State is measured.' While maintaining and defending this position Indonesia has been and is prepared to solve delimitation problems with neighbouring countries through negotiation and agreement (Djalal, 1973a).

The government has sought to negotiate and conclude offshore boundary agreements with its neighbours in order to resolve possible differences before they became political problems. Towards this objective Indonesia has successfully concluded bilateral and tri-partite agreements with most of its neighbouring countries on the boundaries of the continental shelf, as shown in Table 4.4. As stated by the former PERTAMINA President-Director Haryono in Vienna in March 1978: 'Indonesia, recognizing the dominant aspect of ASEAN's configuration being that of ocean rather than land, took the initiative to demarcate offshore boundaries. In most cases boundary demarcations have been settled by agreement among the States usually using the equidistance rule taking the form of a median line while in other cases physiographic features have also been taken into account' (quoted in American Embassy, 1979, p. 19).

Most of the boundary agreements covering the Indonesian continental shelves in the Andaman Sea, the South China Sea, the Arafura Sea and the Timor Sea are now in force. However there are still three outstanding delimitations that have to be settled: that between Indonesia, Malaysia and the Philippines in the Celebes Sea; that between Indonesia and Australia in the Timor Sea; and that between Indonesia and Vietnam in the South China Sea (Wisnoemoerti, 1980). It is expected that difficulties in arriving at an agreement will be less acute in the case of the Indonesia-Malaysia-Philippines continental shelf boundary as the three countries have a close working relationship as members of the five-nation ASEAN (Association of Southeast Asian Nations) than in the case of the Indonesia-Vietnam boundary dispute where the two countries are widely different in their ideological and political make-up.

The Indonesia-Vietnam dispute over the shelf boundaries centres on an area of 28 000 sq. km in the South China Sea where the boundaries drawn up by Indonesia overlap those drawn up by the

TABLE 4.4
AGREEMENTS ON THE BOUNDARIES OF THE CONTINENTAL SHELF BETWEEN INDONESIA AND HER NEIGHBOURS

Signatories	Date Of Signing	Date Of Entering into Force	Boundaries
Indonesia-Malaysia	27 Oct. 1969	11 Nov. 1969	Continental shelves between the two countries.
Indonesia-Thailand	17 Dec. 1972	7 Apr. 1973	Continental shelf boundaries in (i) northern part of Straits of Malacca, and (ii) Andaman Sea.
Indonesia-Malaysia-Thailand	21 Dec. 1972	16 July 1973	Continental shelves in the northern part of Straits of Malacca.
Indonesia-Australia	18 May 1971	18 Nov. 1973	Seabed boundaries in (i) the Arafura Sea and (ii) certain areas off Irian Jaya coasts.
Indonesia-Australia	9 Oct. 1972	8 Nov. 1973	Seabed boundaries in the Timor and Arafura Seas.
Indonesia-India	8 Aug. 1974	17 Dec. 1974	Continental shelf boundaries between the two countries.
Indonesia-India	14 Jan. 1977	15 Aug. 1977	Extensions of the 1974 continental shelf boundaries in the Andaman Sea and the Indian Ocean.
Indonesia-Thailand	11 Dec. 1975	18 Feb. 1978	Seabed boundary in the Andaman Sea.
Indonesia-India-Thailand	22 June 1978	2 Mar. 1979	Trijunction point and delimitation of the related boundaries of the three countries in the Andaman Sea.

Source: Wisnoemoerti, 1980, pp. 9–10.

former South Vietnam government. In line with its archipelagic state concept Indonesia had delineated its continental shelf boundaries in the South China Sea on the basis of the median line between the outermost points of its outermost islands of Anambas and Natuna. The Indonesian claim over the archipelagic waters around the Natuna and Anambas island groups is 'clearly defined, profoundly substantiated and is being maintained through the years' (Djalal, 1979b, p. 43). In 1968 PERTAMINA signed a production-sharing contract with three oil companies—Agip, Tenneco and Phillips—covering the northernmost area of its continental shelf in the South China Sea.

The former South Vietnam government did not delineate its offshore boundaries until 1971, following the enactment of its Petroleum Law in 1970. In 1973 and 1974 it awarded exploration contracts to three oil companies—Shell-BHP, Shell-Cities Service and Sunningdale-Elf—in the southernmost part of its claimed shelf area, which overlapped the areas claimed by Indonesia as part of its continental shelf.

The former South Vietnam government did in fact recognize the Indonesian archipelagic state regime, but was of the view that the location and size of the Indonesian islands in the South China Sea posed 'special circumstances' in the demarcation of the continental shelf boundaries. It therefore held that the median line should be drawn from the coasts of the Vietnamese mainland and from the coasts of the Indonesian mainland (Kalimantan).

Attempts to resolve the problem were complicated by the fact that Indonesia did not have diplomatic relations with the former South Vietnam government. No agreement on the boundaries had been reached when the war ended in Vietnam. The new Government of the Socialist Republic of Vietnam was prepared to negotiate its offshore boundaries with its neighbours. In a joint communique issued on 23 September 1978, signed by Prime Minister Pham Van Dong and President Suharto, the two countries agreed to continue negotiation on the disputed zone (Djalal, 1979b).

In November and December 1979 and in January 1980 PERTAMINA signed five production-sharing agreements with American oil companies, in areas in the South China Sea which are also claimed by Vietnam (see Fig. 4.13). The first contract, signed in

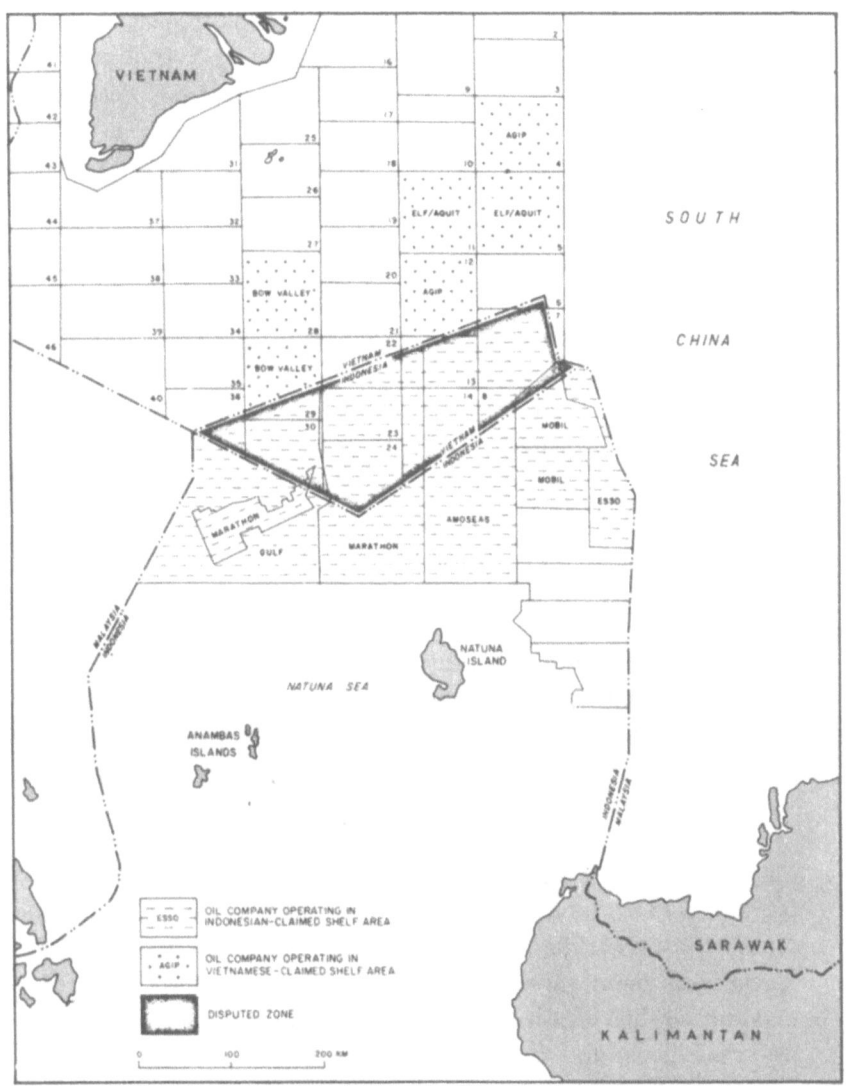

Fig. 4.13   Indonesia–Vietnam disputed zone in the South China Sea
Source: Based on Oil and Gas map, *Petroleum News Southeast Asia*, Jan. 1980.

November 1979 drew a protest from Vietnam. Signing the three December 1979 contracts for Indonesia, Mines and Energy Minister, Dr Subroto stated:

The Natuna area, where explorations are taking place, is Indonesian territory. As a sovereign country, we will protect the safety of foreign companies with all means available . . . Negotiations with Vietnam will continue towards an amicable settlement (*Straits Times*, Singapore, 12 December 1979).

Up to September 1980 Indonesia and Vietnam had held five rounds of discussions on the disputed zone, with Indonesia maintaining that the boundaries should be drawn in accordance with its archipelago state concept but Vietnam holding the position that they should be demarcated following the thalweg concept of the continuous line of greatest depth in the territorial waters between the two countries. According to the Indonesian Foreign Ministry, the talks had been 'very, very difficult' (quoted in the *Straits Times*, 18 September 1980).

The other disputed zone lies in the Sahul Shelf between Indonesia and Australia. The origins of the dispute go back to 1953 when the Australian government claimed the surrounding continental shelf and issued mineral exploration permits in the shelf areas of the Arafura Sea and the Timor Sea. West of longitude 133° 14' east, exploration permits were granted to cover shelf areas up to the southern edge of the Timor Trough. The Australian view was that the Timor Trough, which is a deep depression on the sea floor between 25 and 50 nautical miles from the south coast of Timor and roughly parallel with it, was a natural divide separating the wide Australian continental shelf from the narrow Timor shelf. The Indonesian government did not accede to this view, but held that there was one continental shelf between the two countries, with the Timor Trough as an accidental depression in the sea floor. Such being the case, it wanted the boundary between the two countries drawn on the basis of equidistance between the two shores.

The disputed zone originally covered a large, lens-shaped area (Fig. 4.14). The central portion of this zone covering an area of 9 100 sq. nautical miles was originally the offshore extension of Portuguese Timor. Negotiations between Indonesia and Australia resulted

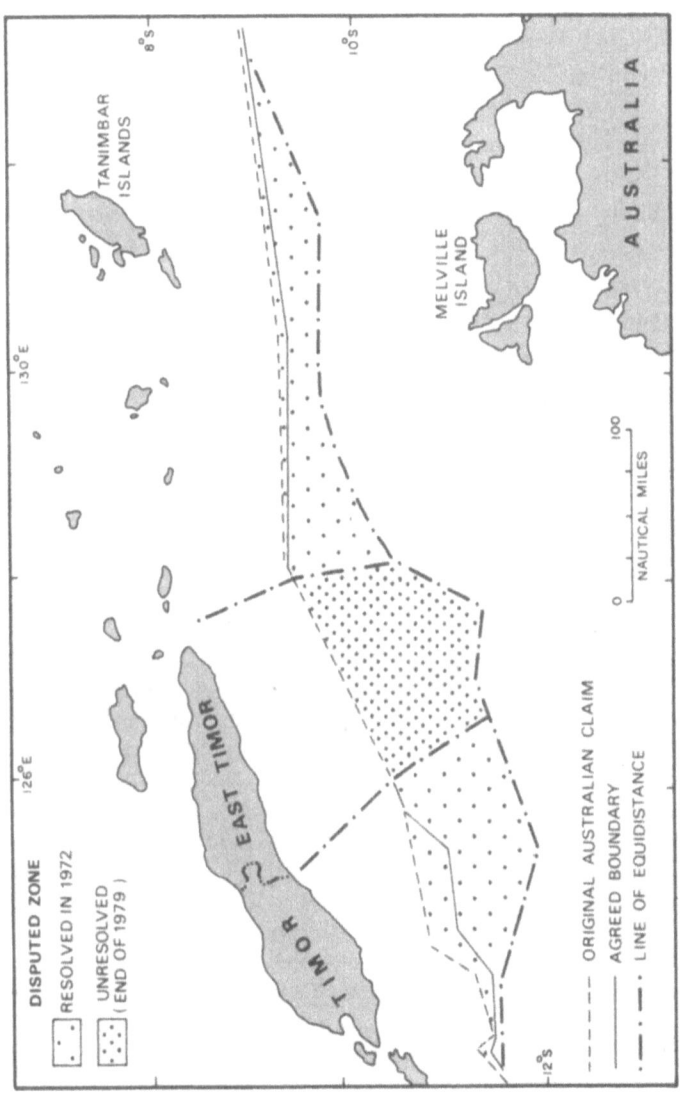

Fig. 4.14 Indonesia–Australia disputed zone in the Timor Sea

Source: Based on Prescott, 1975.

in agreements signed in May 1971 and October 1972 which have resolved the problem of the zone east and west of this central portion, whereby Australia conceded to Indonesia about 750 sq. nautical miles east of the central portion, and about 1 350 sq. nautical miles west of it (Prescott, 1975, pp. 191–4).

The boundary dispute in the Portuguese Timor/Australia continental shelf areas remained unresolved at the time of Portuguese Timor's incorporation by Indonesia in 1976. Indonesia renamed it East Timor, and agreed in 1978 to negotiate a permanent seabed boundary between it and Australia (*Far Eastern Economic Review*, 10 March 1978, p. 79). The second round of talks ended on 27 May 1979 without any decision being made. The settlement of this boundary is more than a matter of academic interest, for the Sunrise and Troubadour gas fields and the prospective kelp structure of Woodside/Arco/Aquitaine would be affected by shifts in the boundary (*Petroleum News*, Supplement, July 1979).

On 21 March 1980 the Indonesian government announced its declaration on the exclusive economic zone of Indonesia, the extent of which is 200 nautical miles measured from the territorial sea baselines.

The Declaration stipulates the sovereign rights and jurisdiction of Indonesia in the exclusive economic zone in accordance with the provisions of the ICNT/Rev.1.[1] It also stipulates that with respect to the seabed and subsoil of the exclusive economic zone, the laws and regulations of Indonesia concerning the Indonesian waters and the Indonesian Continental Shelf, international agreements and international law relating to the continental shelf remain in force. The Declaration also reaffirms that Indonesia will continue to recognize the freedom of navigation and overflight and of the laying of submarine cables and pipelines. With regard to possible overlapping jurisdiction with neighbouring countries, the Indonesian Government is prepared, at an appropriate time, to enter into negotiation with the State concerned with a view to reach an agreement (Wisnoemoerti, 1980, p. 13).

At the time of writing no disputes involving petroleum exploration arising from this Declaration have been reported.

---

[1] Informal Composite Negotiating Text.

V

Production

In 1979 Indonesia produced 2.5 per cent of the total world oil output of 23,653 million barrels. Before the Second World War Indonesia occupied fifth place among the oil producing countries of the world after the United States, the U.S.S.R., Venezuela and Iran. But because no exploration work was conducted during and after the war for a period of almost a quarter of a century, Indonesian oil production from the end of the war until the late 1960s increased at only a very slow rate. Although production subsequently expanded substantially the discovery of large reserves and greatly increased production in the Middle East and elsewhere brought about a relative decline in the position of Indonesia among the world oil producing countries. Thus by 1979 it ranked thirteenth among the world's major producers (Table 5.1), and eighth among the OPEC countries. In 1979 it accounted for only slightly more than 5 per cent of the total OPEC production.

In the context of South-East Asia, however, it is the major oil producer, accounting for about three-quarters of the total production in the region (Table 5.2). The other oil producing countries in South-East Asia are Brunei, Malaysia and Burma, and from 1979, the Philippines. Although an intensive search for oil is being conducted in the other parts of South-East Asia, and oil strikes have been reported in Vietnam (in Thai Binh province, south of Hanoi), and gas finds in the Gulf of Thailand, production has not yet begun in these countries. Available evidence indicates that Indonesia will continue to dominate the oil scene in South-East Asia for the foreseeable future.

Table 5.3 shows the daily average and annual total production of crude oil in Indonesia for the years 1965–79. It will be seen that

TABLE 5.1
WORLD CRUDE OIL PRODUCTION, BY SELECTED
MAJOR PRODUCERS, 1979

Rank	Country	Percentage of World Production of 23,653 Million Barrels
1	U.S.S.R.	18.2
2	United States	15.7
3	Saudi Arabia*	14.7
4	Iraq*	5.3
5	Iran*	4.7
6	Kuwait*	3.9
7	Venezuela*	3.6
8	Nigeria*	3.5
9	People's Republic of China	3.3
10	Libya*	3.2
11	Canada	2.7
12	Mexico	2.6
13	Indonesia*	2.5

Source: Petroleum Economist, January 1980.
*OPEC Member

TABLE 5.2
CRUDE OIL PRODUCTION IN
SOUTH-EAST ASIA, 1979

Country	Barrels per Day Average, First Six Months, 1979	Per Cent
Indonesia	1,606,752	73.8
Malaysia	271,033	12.5
Brunei	255,465	11.7
Burma	30,000	1.4
Philippines[1]	14,100	0.6
Total	2,177,350	100.0

Source: Oil & Gas Journal, 31 December 1979.
[1]On-stream February 1979. Now producing 40,000 b/d.

TABLE 5.3
INDONESIAN CRUDE OIL PRODUCTION, 1965–1979
(in thousand barrels)

Year	Daily Average	Annual Total	Index
1965	485	177,000	100
1966	466	170,000	96
1967	510	186,000	105
1968	603	220,000	124
1969	742	270,000	153
1970	852	311,000	176
1971	890	325,000	184
1972	1,082	395,000	223
1973	1,337	488,000	276
1974	1,373	501,000	283
1975	1,307	477,000	269
1976	1,507	550,000	311
1977	1,685	615,000	347
1978	1,635	597,000	337
1979	1,589	580,000	328

Source: Pertamina Bulletin, February 1978.
1978–9: Direktorat Jendral Minyak dan Gas Bumi, Petroleum & Natural Gas Industry of Indonesia (Jakarta), December 1978 & December 1979.

production had more than trebled during this period. If 1940 is used as a base year then the increase in production between 1940 and 1979 was almost ten times.

Production Patterns

The geographic pattern of production of oil in Indonesia has changed significantly over the years as the older fields ceased to produce or declined in productivity and new fields were discovered and brought on-stream. In the years before the Second World War Kalimantan was the leading producer with 48 per cent of the total Indonesian output of 13 million barrels in 1911, and 68 per cent of the total output of 22.6 million barrels in 1924. The main fields were Sanga-Sanga in the Mahakam River Delta of East Kalimantan, and Tarakan in North-East Kalimantan. Sumatra occupied second position with 42 per cent in 1911 and only 23 per cent in 1924. The main fields were in Aceh (North Sumatra) and in South Sumatra. Java's contribution was 10 per cent in 1911 and 9 per cent in 1924. Irian Jaya remained outside the interests of the oil companies until the 1930s.

The period 1925 to 1968 witnessed the discovery of major onshore oil fields throughout the Indonesian archipelago, but more especially on the Sumatran landmass. Thus of the 24 fields discovered, each with a cumulative production up to the end of 1968 of at least 100 million barrels, 20 were located in Sumatra, 2 in Java, 1 in Kalimantan and 1 in Irian Jaya (Government of Republic of Indonesia, 1972). The most notable find was the giant Minas field in Central Sumatra. Between 1952 and the end of 1968 Minas had produced nearly 1 billion barrels of oil. Other major fields located in Sumatra were Rantau (cumulative production by the end of 1968: 121 million barrels), Duri (170 million barrels), Talang Jimar (144 million barrels), Limau (169 million barrels) and Benakat (107 million barrels). In contrast, none of the fields discovered in Kalimantan, Java and Irian Jaya during this period had a cumulative production of 100 million barrels or more.

As the spate of new discoveries in Sumatra came on-stream this oil-rich island rapidly displaced Kalimantan as the main producing area in Indonesia. The post-war era saw the continued ascendency of Sumatra over the other islands. Throughout the 1950s its share of total production increased yearly; by the 1960s it was responsible for nine-tenths or more of the total Indonesian production. Its percentage share of total production peaked in 1970 when it produced 303 million barrels or 97 per cent of Indonesia's output of 311 million barrels for that year (Table 5.4). Although Sumatra's production continued to increase, reaching a record 406 million barrels in 1973, new onshore finds in Kalimantan, Irian Jaya and Java as well as significant offshore finds in the Java Sea, off East Kalimantan and off Irian Jaya have served to bring about a relative decline in its contribution to total production.

Table 5.4 illustrates the production pattern of the 1970s, specifically, the decreasing share contributed by Sumatra, the increase in the relative outputs of East Indonesia/Irian Jaya and Kalimantan, and more significantly, the steep rise in offshore production.

TABLE 5.4
CRUDE OIL PRODUCTION BY GEOGRAPHIC LOCATION, INDONESIA, 1965–1978

Year	Production (Million Barrels)	Percentage Production by Geographic Location					
		Sumatra	Kalimantan	Java	East Indonesia/ Irian Jaya	Offshore	Total
1965	177	92.9	6.6	0.5	—	—	100.0
1966	170	92.7	6.5	0.4	0.4	—	100.0
1967	186	94.5	4.9	0.3	0.3	—	100.0
1968	220	95.6	3.9	0.2	0.3	—	100.0
1969	270	96.6	3.0	0.2	0.2	—	100.0
1970	311	97.2	2.5	0.1	0.2	—	100.0
1971	325	96.0	2.3	0.2	0.3	1.2	100.0
1972	395	91.7	1.6	0.1	0.1	6.5	100.0
1973	488	83.1	1.4	1.6	0.8	13.1	100.0
1974	501	75.4	1.8	2.4	2.3	18.1	100.0
1975	477	72.1	2.5	1.6	4.9	18.9	100.0
1976	550	62.3	2.9	1.3	5.1	28.4	100.0
1977	615	55.0	2.9	1.2	5.1	35.8	100.0
1978	597	55.1	3.7	1.2	6.6	33.4	100.0

Source: Government of Republic of Indonesia, 1972; Departemen Pertambangan dan Energi, *Buku Tahunan, Pertambangan Indonesia* (Jakarta), various years.

The year 1971 witnessed the first contributions of offshore oil to Indonesian production, with 2,500,000 barrels from the Cinta field north-west of Java and 1,500,000 barrels from the Arjuna field in the Java Sea (see Fig. 5.1). The total amounted to only 1.22 per cent of the Indonesian production for the year, but with the further development of the Cinta/Kitty and Arjuna fields and the coming on-stream of the major fields of Attaka and Bekapai in East Kalimantan production of offshore oil increased to 18 per cent of the Indonesian total in 1974. The mounting importance of offshore oil, a consequence of the concentration of exploration activities in offshore areas in the late 1960s and throughout the 1970s, is reflected in its 28 per cent contribution to total output in 1976 and its 35 per cent contribution to total output in 1977. This percentage declined slightly in 1978.

Production details and the distribution of the offshore fields are shown in Table 5.5 and Figure 5.1 respectively. In 1979 offshore fields contributed 34 per cent of the daily average production for the first six months, and had produced 11 per cent of the total cumulative production of 8.207 billion barrels of oil in Indonesia. Seven of the fields—Handil, Attaka, Arjuna, Bekapai, Cinta, Selatan, Rama—are large and prolific, with daily average production rates ranging from 24,000 to 167,000 barrels. Four of these—Cinta, Arjuna, Attaka, Handil—had by mid-1979 recorded cumulative outputs of over 100 million barrels each, a significant achievement in view of their very recent production history.

The most successful production-sharing companies operating in offshore areas were Total Indonesie/Inpex whose output from their two fields off the Mahakam delta, East Kalimantan, amounted to 39 per cent of the total daily output of Indonesian offshore oil; Union Oil, which produced 22 per cent, mainly from its Attaka field off East Kalimantan; ARCO, which produced 19 per cent, mainly from the Arjuna field in the Java Sea off North-West Java; and IIAPCO, which produced 18 per cent from its seven fields, also in the Java Sea, off North-West Java. CONOCO's Udang field started production in January 1979.

All the offshore oil produced in Indonesia up to 1978 came from fields off the East Kalimantan coast (the offshore extension of the Kutei Tertiary basin) and from fields in the Java Sea, located mainly

TABLE 5.5
CRUDE OIL PRODUCTION FROM OFFSHORE FIELDS, INDONESIA, 1979

Name of Company	Name of Field	Discovery Date	Daily Average Production, First 6 Months, 1979 (Barrels)	Per Cent	Cumulative Production to Mid-1979	Per Cent
IIAPCO	Cinta	1970	32,235	5.9	116,056,060	12.5
	Kitty	1973	3,670	0.7	8,533,472	0.9
	Nora	1973	1,471	0.3	5,785,440	0.6
	Rama	1974	24,327	4.5	49,082,335	5.3
	Zelda	1971	1,033	0.2	2,666,402	0.3
	Selatan	1971	30,292	5.6	5,844,107	0.6
	Gita	1972	3,209	0.6	580,830	0.1
ARCO	Arjuna	1969	98,691	18.1	231,997,582	25.0
	Arimbi	1972	7,040	1.3	11,005,381	1.2
Union Oil	Attaka (with Inpex)	1970	102,237	18.8	227,907,518	24.6
	Melahin	1972	1,834	0.3	2,955,453	0.3
	Sepinggan	1973	9,873	1.8	21,143,282	2.3
	Kerindingan	1972	1,428	0.3	1,170,780	0.1
	Yakin	1976	2,660	0.5	1,851,692	0.2
Total Indonesie/ Inpex	Bekapai	1972	47,649	8.7	69,053,658	7.4
CONOCO Group	Handil	1974	166,912	30.6	170,622,934	18.4
	Udang	1974	9,825	1.8	1,778,301	0.2
Total	17 fields		544,386	100.0	928,035,227	100.0

Source: Data from *Oil & Gas Journal*, 31 December 1979.

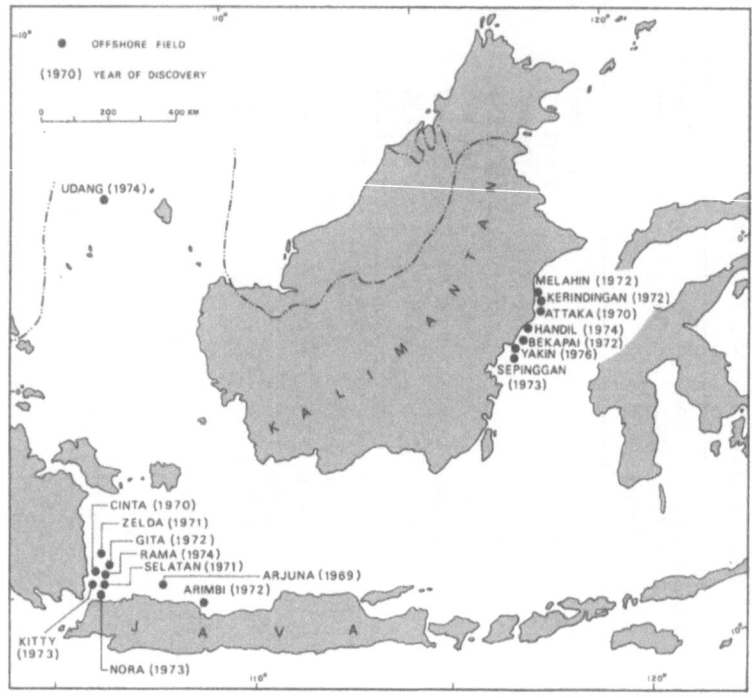

Fig. 5.1   Distribution of offshore oil fields in Indonesia

in the North-West Java Tertiary basin. In 1979 a new offshore area
was brought into the picture, with the coming on-stream of the
Udang field in the Natuna Sea (Fig. 5.1).

The production pattern by companies operating under different
contractual arrangements is shown in Table 5.6. Certain trends are
evident: first, actual production by government-run PERTAMINA
and Lemigas almost doubled during the period 1970—9 but their
percentage share of total production declined due to the substantial
increase in total Indonesian production. Second, actual production
by companies operating under Contracts of Work remained static
between 1970 and 1979 but similarly, their percentage share of total
production declined from 88.5 to 48 per cent during this period. It is
likely that in the 1980s production from such companies will cease as
the Government of Indonesia has announced its intention of either
terminating all Contracts of Work arrangements in 1983 or

TABLE 5.6
CRUDE OIL PRODUCTION BY FORMS OF CONTRACT, INDONESIA, 1970−1979

Year	Total Production (Million Barrels)	Percentage Production			
		PERTAMINA & Lemigas	Contract of Work Companies	Production-Sharing Companies	Total
1970	311	10.3	88.5	1.2	100.0
1971	325	10.1	87.7	2.2	100.0
1972	395	7.9	83.7	8.4	100.0
1973	488	7.6	76.8	15.6	100.0
1974	502	8.3	69.4	22.3	100.0
1975	477	6.9	66.4	26.7	100.0
1976	550	5.8	58.2	36.0	100.0
1977	615	6.0	49.7	44.3	100.0
1978	597	5.3	48.5	46.2	100.0
1979	580	5.3	48.0	46.7	100.0

Source: Direktorat Jendral Minyak dan Gas Bumi, *Petroleum & Natural Gas Industry of Indonesia* (Jakarta), various years.

converting them to production-sharing contracts. Third, both the actual and percentage share of production by production-sharing companies increased significantly, from 3.8 million barrels (1.2 per cent) in 1970 to 271 million barrels (47 per cent) during the period under review. Again, consequent upon the Indonesian government's declared intention to terminate or convert all Contracts of Work arrangements to production-sharing contracts, it is likely that all production in the latter part of the 1980s, apart from that from PERTAMINA and Lemigas, will be from companies operating under production-sharing contracts.

The production of individual oil companies in 1979 is summarized in Table 5.7. The dominance of Caltex, producing oil under a Contract of Work, is apparent. This dominance is derived mainly from the company's discovery of the bonanza 30,000 ha Minas field, which to date has produced more than two billion barrels of oil or over 30 per cent of the total cumulative production in Indonesia since production of oil first began in 1885. It is currently still producing 337,000 barrels

TABLE 5.7

INDONESIAN CRUDE OIL PRODUCTION BY COMPANY, 1970 AND 1979

Company	Production, 1979 (Barrels)	Per Cent	Production, 1970 (Barrels)	Per Cent of Total	Per Cent Change in Production, 1970 adn 1979
Caltex*	266,047,600	45.8	257,877,300	82.8	+ 3
Inpex (Japex)**	57,180,500	9.9	Production started in 1973		
ARCO**	41,324,700	7.1	Production started in 1971		
Total Indonesie**	38,691,200	6.7	Production started in 1974		
IIAPCO**	32,730,700	5.6	Production started in 1971		
PERTAMINA	30,315,900	5.2	31,760,400	10.2	− 5
Union Oil**	23,968,800	4.1	Production started in 1972		
Petromer Trend**	22,662,500	3.9	Production started in 1973		
Mobil Oil**	20,789,000	3.6	Production started in 1976		
Calasiatic & Topco*&***	12,534,900	2.2	Production started in 1973		
Stanvac*	10,810,900	1.9	17,674,400	5.7	− 39
Roy M. Huffington**	7,663,100	1.3	Production started in 1974		
CONOCO**	6,184,000	1.1	Production started in 1979		
Phillips**	3,105,100	0.5	Production started in 1975		
Tesoro**	3,060,300	0.5	Production started in 1972		
Asamera**	2,835,600	0.5	3,775,100	1.2	− 25
AARNL**	329,100)		Production started in 1971		
Lemigas	212,700)	0.1	464,700	0.1	− 54
Total	580,446,600	100.0	311,551,900	100.0	+ 86

*Contract of Work
**Production-sharing

Source: Data from Direktorat Jendral Minyak dan Gas Bumi, *Petroleum and Natural Gas Industry of Indonesia* (Jakarta), January−December 1979.

a day. In addition Caltex's concession area in Central Sumatra contains thirty-seven other fields, and several more have been discovered recently. As Minas and the other fields came on-stream, production increased rapidly so that by 1970 Caltex alone was responsible for four-fifths of the total Indonesian oil output. Between then and 1979 the company's production increased only slightly from 258 million barrels to 266 million barrels. Its share of total Indonesian output consequently declined to 46 per cent because of the substantial contributions made by the production-sharing companies. The dominant position of Caltex is likely to continue into the future as the Minas field still appears to contain significantly large quantities of oil; as the company has discovered and is working new fields within the concession area as well as in holdings outside the area held under a production-sharing contract; as the company has embarked on a secondary recovery effort to enhance production in its existing fields; and as the Indonesian government had, in 1978, announced the extension of its present contract with Caltex to the year 2013.

The position of Stanvac, the other major Contract of Work company, stands in contrast to that of Caltex, in that not only is its production a small fraction of that of the latter, but production had also declined between 1970 and 1979.

Although none of the other (production-sharing) companies can match Caltex in production, several of them have, in a relatively short period, increased their oil output considerably. Notable among these are Inpex (JAPEX), with its oil fields in East Kalimantan; Atlantic Richfield Indonesia (ARCO) which has its oil fields, Arjuna and Cinta, offshore to the north-east of Jakarta Bay; Total Indonesie, with its onshore field in the Mahakam delta and offshore fields, Bekapai and Handil located off East Kalimantan; IIAPCO, the first company to produce oil offshore, with its main fields, Cinta, Kitty, Selatan, Nora, Zelda, Rama and Gita, located west of North-West Java; Union Oil with offshore fields, Attaka, Melahin, Kerindingan, Sepinggan and Yakin, in East Kalimantan, and Petromer Trend, with oil fields, Kasim, Jaya, and Walio, located in the Vogelkop area of Irian Jaya. All of these companies produced 4 per cent or more of the total Indonesian output in 1979. PERTAMINA, the state enterprise, occupied sixth position in 1979, with 5 per cent of the total production. By 1979 Inpex had become the largest production-sharing producer, with 58 million barrels (10 per cent). Occupying lower positions were ARCO with 41 million barrels (7 per cent), Total Indonesie with 39 million barrels (7 per cent) and IIAPCO with 33 million barrels (6 per cent).

Distribution of Oil Fields

There are over 90 known petroleum accumulations in Indonesia, of which 3 are in the 'very large' category, and 41 are accumulations with between 50 and 500 million barrels of oil (Table 5.8).

At the beginning of 1976 Indonesia had a total of 237 oil fields of various sizes (Table 5.9). These fields were located in all the main islands in the Indonesian archipelago as well as in the offshore areas (Fig. 5.2). Most of these fields are small, with daily average production ranging from a few hundred to a few thousand barrels. In their analysis of 120 oil and gas fields by field size, Soeparjadi *et al.* (1975) noted that there were 5 fields of super-giant size (over 1 billion barrels discovered reserves), 1 field of giant size (500–1,000 million barrels), 34 fields of major size (50–500 million barrels), and 80 fields of minor size (1–50 million barrels).

In mid-1979 there were 115 oil fields listed as producing fields, excluding a number of very small fields. Table 5.10 shows the size distribution of these fields. A number of significant features are revealed in the Table. First, six fields—Minas, Handil, Attaka, Arjuna, Arun and Bangko—among them produced exactly half of the

TABLE 5.8
KNOWN PETROLEUM ACCUMULATIONS
IN INDONESIA

Size	Accumulations in Million Barrels	Number
Very large	Over 500	3
Large	100 to 500	15
Fairly large	50 to 100	26
Small	<50	>50

Source: Departemen Pertambangan, *Buku Tahunan Pertambangan Indonesia*, 1978, p. 45.

TABLE 5.9
OIL FIELDS IN INDONESIA, 1976

Location	Number of Fields
Sumatra	
North Sumatra	22
Central Sumatra/Riau	53
South Sumatra/Jambi	75
Java	12
East Kalimantan	56
South China Sea	2
Natuna Islands	1
Irian Jaya	14
Seram Island	2
Total	237

Source: PERTAMINA.

total Indonesian output of 1,606,700 barrels a day. Second, at the other end of the scale, 54 fields individually producing less than 4,000 barrels a day, plus a number of unlisted small fields, together making up more than half of the oil fields, were responsible for only 7 per cent of the daily production in mid-1979. Third, more than two-thirds of the oilfields, each producing less than 6,000 barrels a day, were collectively responsible for only 11 per cent of the total daily production. These figures provide statistical confirmation that most of the Indonesian oil fields are small and not particularly prolific.

However some are of significant size and productivity and may be considered major fields. A major field is defined here as being either (1) a field with a cumulative production of 100 million barrels or more or (2) a field with a daily production of 20,000 barrels or more. Figure 5.3 shows the major fields ranked by cumulative production, while Figure 5.4 shows the major fields ranked by daily average production. They reveal certain aspects of petroleum production in Indonesia:

1. The cumulative production of all the oil fields in Indonesia up to mid-1979 was 8.207 billion barrels. The cumulative production of all the oil-producing countries in South-East Asia was 10.616 billion barrels. In the South-East Asian context Indonesia is therefore by far

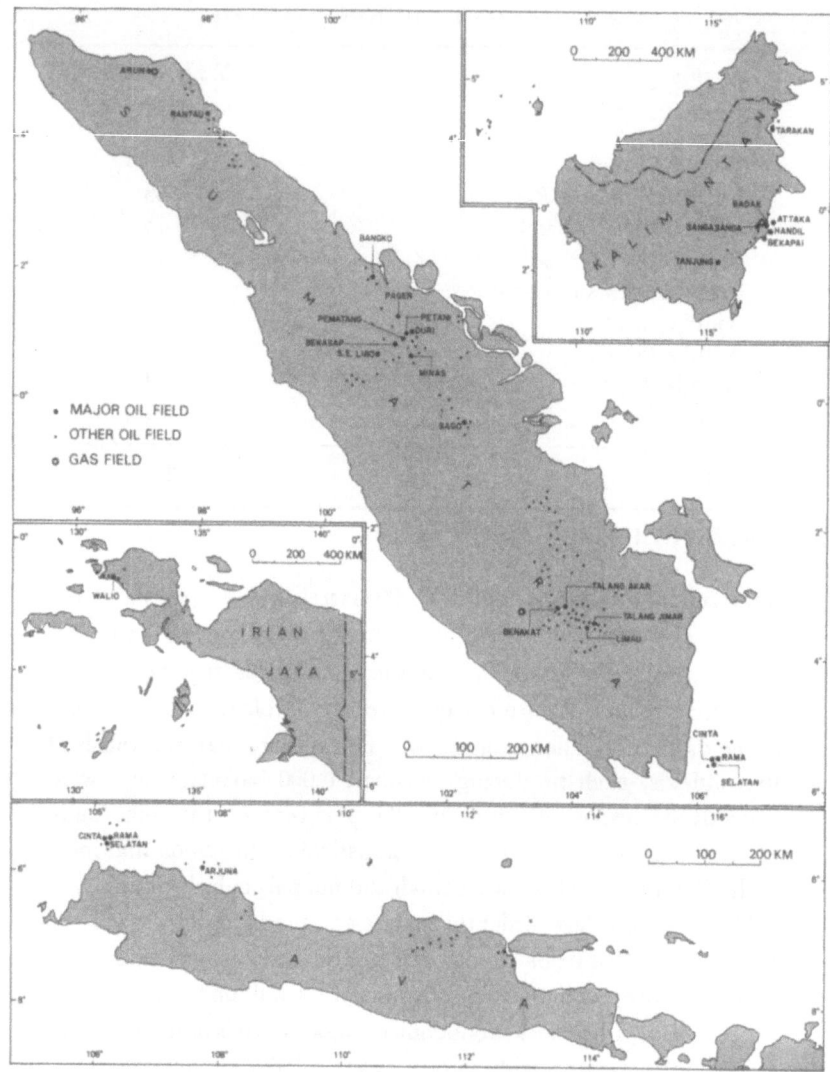

Fig. 5.2 Distribution of the oil and gas fields in Indonesia

TABLE 5.10
SIZE DISTRIBUTION OF PRODUCING FIELDS
IN INDONESIA, 1979

Size (Barrels/Day)	Number of Fields	Per Cent	Total Barrels/Day during First Six Months, 1979
Over 200,000	1	1	337,134
100,000 − 199,999	2	2	269,149
50,000 − 99,999	3	3	204,066
40,000 − 49,999	2	2	91,421
30,000 − 39,999	3	3	98,602
20,000 − 29,999	7	6	170,735
10,000 − 19,999	10	8.5	151,057
9,000 − 9,999	6	5	57,120
8,000 − 8,999	2	2	17,149
7,000 − 7,999	3	3	22,305
6,000 − 6,999	2	2	13,039
5,000 − 5,999	8	7	44,220
4,000 − 4,999	4	3	18,625
3,000 − 3,999	11	9	38,247
2,000 − 2,999	10	8.5	23,605
1,000 − 1,999	19	16	25,713
100 − 999	22	19	15,114

Other small fields in PERTAMINA, Lemigas, Caltex, Stanvac, Asamera, Huffco, Tesoro and Petromer Trend areas.

Source: Compiled from *Oil & Gas Journal*, 31 December 1979.

the most important oil-producer, with 77.3 per cent of the cumulative‧ production. Brunei, the next ranking country, produced 14 per cent, Malaysia 4.7 per cent and Burma 3.9 per cent. Thailand has recorded a cumulative production of only 1 million barrels (1978), while production from the Philippines has only just begun, and stood at 2 million barrels (*Oil & Gas Journal*, 31 December 1979).

2. Major fields, as defined above, span the entire production history of Indonesia. One or more major oil fields have been discovered in each decade, except the 1910s, since oil was first produced in Indonesia in the 1890s. The 1970s may be regarded as the decade of offshore production as half of the twelve major fields discovered in this decade were offshore fields.

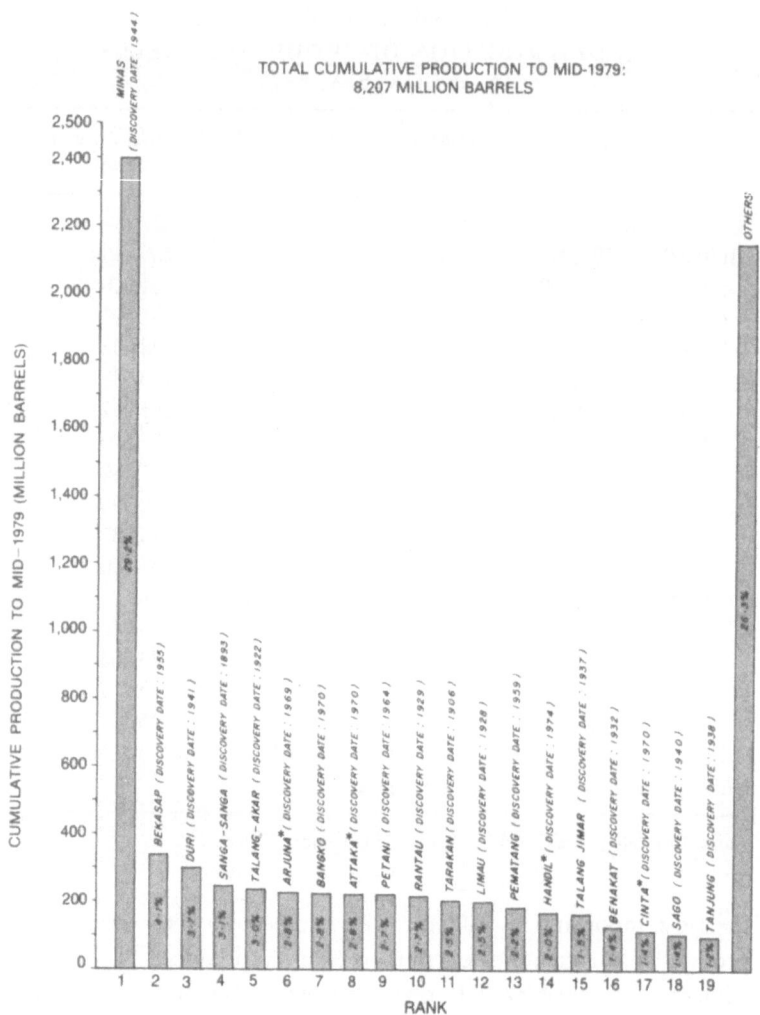

Fig. 5.3   Major oil fields by cumulative production, 1979
Source: Data from *Oil & Gas Journal*, 31 December 1979
*Offshore

3. Ten of the fields shown in Figures 5.3 and 5.4 are major fields as defined in terms of both cumulative production as well as daily average production. These fields are: Minas, Handil, Arjuna, Attaka, Bangko, Cinta, Duri, Petani, Bekasap and Pematang (Fig. 5.5). Four of them—Handil, Arjuna, Cinta, Attaka—are offshore fields.

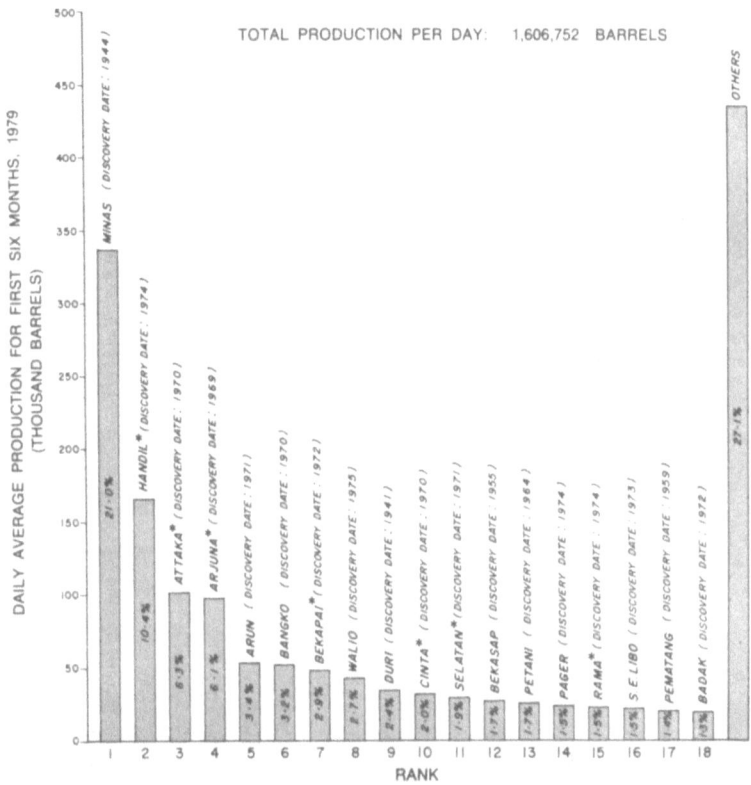

Fig. 5.4   Major oil fields by daily average production, 1979
Source: Data from *Oil & Gas Journal*, 31 December 1979
*Offshore

4. Minas is by far the largest field ever discovered in Indonesia. Although it was discovered only in 1944 it has produced 29 per cent of the total cumulative production, and in mid-1979 was still responsible for 21 per cent of the daily average production of 1.607 million barrels. The fact that the next largest field (Bekasap) in terms of cumulative production accounted for only 4 per cent of the total production is a good indication of the relative size and productivity of Minas.

5. Although the history of offshore production began only in 1971, by mid-1979 seven of the major fields defined by daily average

140

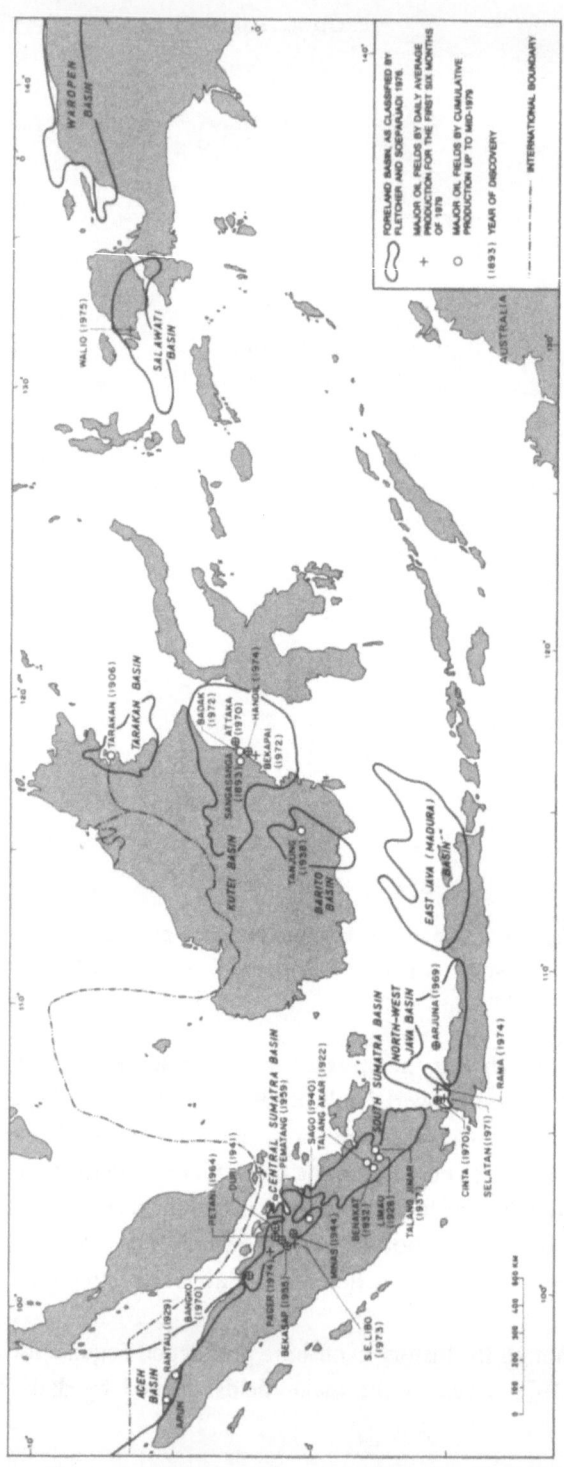

Fig. 5.5 Distribution of the major oil fields in Indonesia

production were offshore fields. Together these seven majors accounted for 31 per cent of the daily average production. Three of these—Handil, Attaka, Arjuna—are highly prolific fields, with daily average production of 100,000 barrels or more. Handil's production was half that of Minas (Fig. 5.4). Four of the offshore fields have reached major-field status as defined by cumulative production. These fields—Arjuna, Attaka, Handil, Cinta—together accounted for 9.1 per cent of Indonesia's total cumulative production of 8.207 billion barrels in mid-1979 (Fig. 5.3).

As Table 5.11 and Figure 5.5 show, all the major oil fields in Indonesia, onshore as well as offshore, are located in foreland basins of Tertiary age. The next section is devoted to brief descriptions of some of the fields in these basins.

OIL FIELDS IN THE ACEH BASIN

The Aceh (North Sumatra) basin extends from the north-eastern tip of Sumatra into the Straits of Malacca. The production history of the basin goes back to the 1880s when the first oil well at Telaga Said was developed. Between then and 1968 the basin produced over 200

TABLE 5.11
BASINAL CHARACTERISTICS OF THE MAJOR
OIL FIELDS IN INDONESIA

Name of Field	Name of Tertiary Basin	Basin Type [1]
Rantau, Arun	Aceh (North Sumatra)	Foreland
Minas, Bekasap, Duri, Petani, Bangko, Pematang, Pager, Sago, S.E. Libo	Central Sumatra	Foreland
Talang Jimar, Benakat, Limau, Talang-Akar	South Sumatra	Foreland
Arjuna, Cinta, Rama, Selatan	North-West Java	Foreland
Tarakan	Tarakan (Kalimantan)	Foreland
Sanga-Sanga, Attaka, Handil, Bekapai, Badak	Kutei (Kalimantan)	Foreland
Tanjung	Barito (Kalimantan)	Foreland
Walio	Salawati (Irian Jaya)	Foreland

[1] As defined by Fletcher & Soeparjadi (1976).

million barrels of oil. The oil fields were originally worked by Shell/ BPM but were bought over by the Indonesian government and are now operated by PERTAMINA.

The productive zones are the shallow Baong and post-Baong sands of Upper Miocene and Early Pliocene age, found in gently folded anticlines located in a narrow strip along the north-eastern coast. Of the 37 hydrocarbon accumulations discovered in these reservoirs only 12 are still producing minor amounts of oil (Kingston, 1978). In fact the sandstones of the middle part of the Baong Formation have been known to be oil-bearing since the discovery of the Telaga Said and Darat fields at the beginning of the production history of Indonesia, more than seventy-five years ago (Mulhadiono *et al.*, 1978). The most important field in this basin is Rantau, discovered in 1929. Faulting in the Rantau anticlinal structure has divided it into various blocks with the hydrocarbons present at different structural levels. More than twenty productive zones have been discovered at levels varying from 300 to 1 400 m. The crude is light paraffinic with an average gravity of 48.5 ° API. Cumulative production up to mid-1979 was over 219 million barrels. Production in this relatively old field has declined in recent years, from 13.2 million barrels in 1972 to 9.7 million barrels in 1975. Average daily production has also declined, from 14,470 barrels for the first six months of 1977 to 9,293 barrels for the first six months of 1979 (*Oil & Gas Journal*, 31 December 1979).

Several thousand m below the oil-producing sands of the Aceh basin are the calcareous shales of the Peutu Formation, containing carbonate reefs such as the Arun Limestone. These reefs of Lower Miocene age have been proved, in the centre portion of the basin, to hold the largest hydrocarbon accumulation in the basin, in the form of non-associated gas estimated to total 17 trillion cu. ft or the equivalent of 3 billion barrels of oil. The gas in this field—the Arun gas field—occurs at a depth of about 3 000 m. The field itself is oval-shaped and is 18 km long by 5 km wide, and has an average thickness of 153 m. Both reservoir pressure and temperature are unusually high, at 7,100 lb/sq. in and 178 °C (352 °F) respectively. The Arun field has been producing 4.5 million metric tons of LNG a year since 1977, in a plant which is now the fourth largest in the world and capable of further expansion. The Arun gas also contains commercial

quantities of condensate (a heavier hydrocarbon), and the LNG plant
contains facilities for processing, storing and shipping the liquid.

OIL FIELDS IN THE CENTRAL SUMATRAN BASIN

By far the largest and most prolific field is Minas (Fig. 5.6),
classified as a giant oil field in petroleum circles. The field is located in
the Central Sumatra Tertiary foreland basin which experienced a

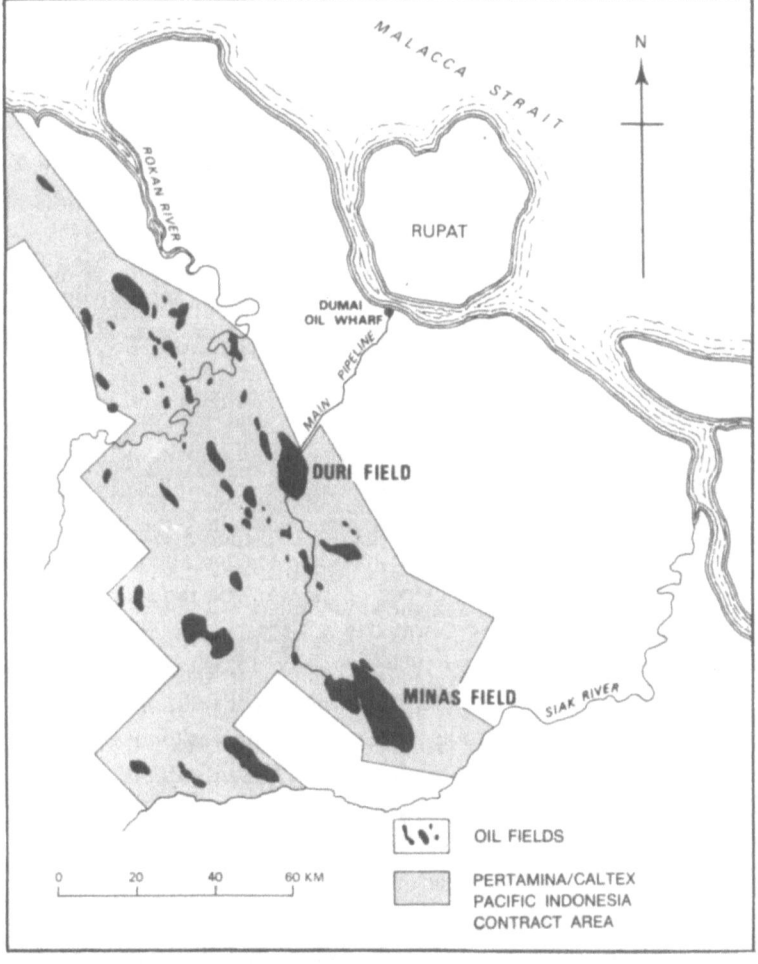

Fig. 5.6 The Minas and Duri oil fields
Source: Based on Hasan, *et al.*, 1977/8.

transgressive-regressive cycle of deposition. During the Early Miocene deep marine shales (Telisa Formation), rich in organic matter, were deposited to the west of the basin. They form the major source of hydrocarbons. To the east and north-east shallow water shelf facies (Sihapas Group) were laid down to form a reservoir system for the accumulation of hydrocarbons from the Telisa Formation through lateral migration (Mertosono & Nayoan, 1974).

The Minas anticline is a very large gently folded anticline, about 45 km long by 24 km wide. The north-west part of the structure is cut by the Minas fault, a major strike slip fault. The productive part of the field is 24 to 28 km long and 7 to 10 km wide. The oil column is 130 m. As in the other parts of the Central Sumatra basin, the Sihapas Group, a sequence of sand and shale about 180 m thick, is the main producing interval, and in the Minas field is found at depths of between 600 to 800 m. The Sihapas interval consists of three-quarters sand and one-quarter shale. Oil from the Minas field is from five major transgressive sand units, and in fact the sands account for 80 to 90 per cent of the oil produced in the Caltex contract area of Central Sumatra.

Early exploration work on Minas started in 1938, and by 1940 a seismic survey had confirmed the existence of a gently folded, very large, closed anticline, holding promise of a major oil accumulation within it. The discovery well was abandoned when the Pacific War broke out, and it was left to the Japanese to spud and complete the well in December 1944. The end of the war saw the resumption of drilling (in 1950) which confirmed that Minas was indeed a major oil field. Production began in 1952 averaging 15,000 barrels per day. Additional extension and stepout development wells drilled proved the existence of substantially more productive areas to the south-east and north-west of the field (Hopper, 1976; Hasan, *et al.*, 1977/8).

Production increased steadily and by 1972/3 Minas was producing at an average rate of over 400,000 barrels of oil per day. Since then daily average production has dropped, to 337,000 barrels per day in the first six months of 1979. Minas has recorded a cumulative production of 2,394 million barrels of oil by mid-1979, or 29 per cent of the total Indonesian cumulative oil production, from 1885 to mid-1979, of 8,207 million barrels. Caltex has put the oil-in-place at

7 billion barrels (*Petroleum News*, January 1980). The estimated ultimate recoverable reserves are over 4 billion barrels (M. T. Halbouty, quoted in Beddoes, January 1980), so that over 1 billion barrels remain to be lifted.

The company initiated a secondary recovery programme at Minas in 1970, involving the drilling of water injection and production wells in and around the existing wells. Several hundred thousand barrels of water are being injected per day into the Minas producing formations to boost the underground pressure and drive the oil towards the wells.

The Minas crude is used as a 'marker' crude in Indonesia and has become recognized in the world oil markets as a waxy, low sulphur oil. It is paraffinic, 35.8 ° API gravity, and has only 0.08 per cent sulphur by weight.

Within the same Tertiary basin are two other large fields: Duri and Bangko, both in the Caltex area of operations. The Duri field (Fig. 5.6) is in a large anticline cut by a number of small faults. The productive section of the anticline is about 17 km long by 8 km wide, covering an area of about 9 900 ha. As in the case of the Minas oil field, the oil from the Duri field is from the Sihapas Group reservoir sands. The Sihapas Group represents the transgressive, coarse clastic sequence deposited during the Early Miocene. The uppermost unit of this Group consists of sandstones and shales of the Duri Formation. The hydrocarbons found in the Duri Formation were probably sourced from the laterally equivalent Telisa Formation shales (Wongsosantiko, 1976). The structure of the field is a large anticline with its axis trending north-south. Caltex has discovered twelve oil and gas fields in Duri Formation reservoir sands, of which eight are in production.

The Duri field, discovered in 1941, is known to contain 6 billion barrels of oil-in-place, but for technical reasons primary recovery methods will yield only 7.4 per cent of this oil. As only 300 million barrels of oil have been recovered from this field since it started production, some 5.7 billion barrels of the oil-in-place remain untapped. In order to ascertain the economic feasibility of using enhanced secondary recovery techniques to increase the percentage recovery from the field, Caltex started a field test of steam injection in a test area in 1975. Between 1975 and 1978 it injected over 12 million barrels of water in the form of steam into the sixteen wells of the test area. About 1.75 million barrels of additional oil was produced as a result of this test (Mathews, 1979).

The Duri field has recorded a cumulative production of nearly 303 million barrels by mid-1979. Daily average production increased slightly to 36,075 barrels in the first six months of 1979 as compared to 35,098 barrels over the same period in 1978 (*Oil & Gas Journal*, 31 December 1979).

Apart from Minas and Duri, Caltex operates fifty-one other oil fields of varying sizes and productivities in the Central Sumatra basin. The largest of these is the Bangko field. Discovered in 1970 Bangko has within a short period recorded a cumulative production of 321 million barrels by mid-1979, and in the first six months of 1979 was producing a daily average of 50,979 barrels of oil. Three other large fields in the Caltex area are Bekasap, discovered in 1955, with a cumulative production of 340 million barrels by mid-1979, and a daily average production of 27,828 barrels in the first six months of 1979; Pedada, discovered in 1973, producing 16,760 barrels per day in the first six months of 1979; and Balam South, discovered in 1969, and producing 16,586 barrels per day.

The other major oil company operating in the Central Sumatra basin is Stanvac. It had ten fields in the basin in 1979, of which eight were in production. The largest of these was Sago, discovered in 1940, with a cumulative production of 113 million barrels up to mid-1979, and a daily average production of 3,653 barrels of oil in the first six months of 1979.

OIL FIELDS IN THE SOUTH SUMATRA BASIN

This basin has been more thoroughly explored than any other basin in Indonesia. Some thirty-five oil fields were discovered in the basin between 1898 and 1958 but several of these have since been depleted. Although South Sumatra had the largest number of fields in Indonesia in 1976 (Table 5.9), production from this area declined steadily, from 30.2 million barrels in 1972 to only 16.4 million barrels in 1976 (Departemen Pertambangan, Indonesia, 1976). The largest of the fields in this basin is the Talang-Akar—Pendopo field, which from the year (1922) of its discovery by Stanvac to 1968 produced more

than 337 million barrels of paraffinic low sulphur oil of 32—37 ° API
gravity. The oil accumulations are developed in a gentle
asymmetrical anticline cut by several cross-faults at right angles to
the structural axis. The more prolific Talang Akar culmination lies to
the north, while Pendopo lies to the south. Production is from the
deeper prolific sands of the Lower Telisa Formation of Lower
Miocene age. The productive area is about 19 km long and 3 km wide.
The reservoir and structural characteristics of the field are such that
gas injection had to be applied from as early on as 1927 to maintain
reservoir pressure and allow for higher recoveries of oil.

Two fields discovered recently in the South Sumatra basin are
Belimbing, which lies to the west of Pendopo, and East Benakat,
located east of Talang Akar. The Belimbing field was discovered by
PERTAMINA in 1965 and had produced 18.4 million barrels of oil
by mid-1979. Production has declined considerably in recent years,
and was only 780 barrels per day in mid-1979. The East Benakat
field was also discovered by PERTAMINA when it drilled
exploration wells in the area in 1972 and located oil in commercial
quantities in the Talang-Akar Formation. The main structural
element of the field is an elongated structural high 12 km long by 1.5
km wide (Akuanbatin & Ardiputra, 1976). Production from this field
was 5,000 barrels per day in mid-1979.

OIL FIELDS IN THE NORTH-WEST JAVA BASIN

The major fields in this basin include the onshore fields of
Jatibarang/Bongas and the offshore fields of Arimbi and a cluster of
fields in the Java Sea: Arjuna, Gita, Rama, Selatan, Nora, Zelda,
Cinta, and Kitty. The Bongas field in West Java was discovered in
1940, and recorded a cumulative production of only 310,000 barrels
as at the end of 1968. It has ceased to be of significance today.
Jatibarang, however, is a recently discovered field operated by
PERTAMINA. It is located some 170 km east of Jakarta and named
after the town of Jatibarang which lies 20 km west of it. Seismic
surveys in the area were started in 1967 and an exploratory well
drilled in 1969 discovered oil at a depth of 2 010 m. The crude oil from
this field is highly waxy, sulphur-free, with a gravity of 30° API, and a
high pour-point of 110 °F to 115 °F. For the first few years after its

discovery production was only slightly over 20,000 barrels per annum. In 1973, however, production increased to 7.3 million barrels, and in 1974 to 11.5 million barrels. Production the following year dropped to 5.8 million barrels. By mid-1979 it had a cumulative production of nearly 50 million barrels, and a daily output of 18,238 barrels. Production in the Jatibarang field is from thirty-eight wells between 1 980 m and 2 500 m deep. Reserves in this field are estimated to be about 160 million barrels (Soejoso & Indrarto, 1975).

The North-West Java offshore fields (Fig. 5.1) are part of a new petroleum province discovered in the Java Sea in the late 1960s and early 1970s. The contractors to this area are the IIAPCO and ARCO groups. Production-sharing contracts signed with PERTAMINA in 1966 was followed by seismic surveys in 1967. Exploratory drilling began in September 1968. In the Offshore Northwest Java Contract Area operated by the ARCO group the first commercial discovery within the Arjuna complex was made in December 1968. Between then and the end of 1973 a total of thirty-two exploratory wells were drilled resulting in seven discoveries. Production started in August 1971 at the rate of 13,000 barrels per day from the 'E' structure of the field. As this was the first time that oil in commercial quantities had flowed from an offshore field, it marked the beginning of a new era in the history of petroleum development in Indonesia, an era that has become increasingly important as more and more offshore discoveries come on-stream.

Further development of the structures in the Arjuna complex resulted in a substantial increase in production from this field. By the end of 1973 a total of fifty-four wells had been drilled from ten platforms, and production had increased to 120,000 barrels per day. Cumulative production from this field by mid-1979 was 232 million barrels, and daily production 98,691 barrels.

Although oil has been discovered in sediments of Oligocene/Miocene age, most of the production from Arjuna is from the Upper Cibulakan sands of Miocene age. These were deposited in low energy, shallow water, open marine conditions in a stable shelf area. The crude is typically low sulphur (0.08 per cent by weight), with a gravity of 37 ° API, a pour-point of 80 °F, and a wax content of 12 per cent (Ray, 1974; Scheidecker & Taiclet, 1976).

The Cinta field lies in the Southeast Sumatra Contract Area operated by the IIAPCO group. The field was discovered in September 1970 and began production the following year. Production is from the Batu Raja limestone at a depth of 850 m and from the Talang-Akar sandstone at a depth of 1 000 m. The crude from this field has a gravity of 34.5° API, a wax content of 26.6 per cent by weight, a sulphur content of 0.07 per cent by weight, and a high pour-point of 105 °F. Cumulative production from 1971 to mid-1979 was an impressive 116 million barrels, with a daily output of 32,235 barrels. The favourable geologic and operating conditions point towards a rising production rate for some years to come (Nadom & Ramsay, 1972). Subsequent discoveries in the area—Selatan and Zelda (1971), Gita (1972), Kitty and Nora (1973), and Rama (1974)—have led to a significant increase in production from this sub-basin. Rama has recorded a cumulative production of 49 million barrels and was producing 24,327 barrels a day by mid-1979. Average daily output from Selatan was 30,292 barrels during the same period; the other four fields together produced 9,383 barrels per day.

OIL FIELDS IN THE BARITO BASIN

Only one field of significant size has been found in the Barito basin. This is the Tanjung field. Exploration started in 1930 and the first discovery that there were commercial quantities of oil in the field was made in 1938. However production did not start until 1962, after the completion of a pipeline to Balikpapan. The field is an elongated anticline with a NNE to SSW axis, bounded by a transverse fault to the west. The Tanjung Formation in which the oil is found is the oldest of the Tertiary sediments in the Barito basin. The oil is in sands belonging to the Lower Eocene basal conglomerate zone lying below a 30 m thick dolerite sill intruded between the bedding planes of the sediments. The five payzones are located at approximately 100 m intervals between the 710 and 1 015 m zones (Sinegar & Sunaryo, 1980). Cumulative production up to mid-1979 was 101 million barrels. Average daily production for the first six months of 1979 was 5,121 barrels. The oil is paraffinic, with a sulphur content of 1 per cent by weight, and a gravity of 39.7° API. Much of the Barito basin remains to be explored. The potential appears to be good, as source rocks are abundant in this foreland basin (Fletcher & Soeparjadi, 1976).

OIL FIELDS IN THE KUTEI BASIN

The Kutei Tertiary foreland basin of East Kalimantan is a very extensive basin which has been spasmodically explored for oil since the late nineteenth century. A large field, Sanga-Sanga, was discovered in 1898 and has been delineated by over 900 wells since then. The Sanga-Sanga structure is a markedly asymmetrical anticline over 32 km long and aligned in a NNE-SSW direction. The field is split in two by the Mahakam River, and is seldom more than 1 km wide. Production is from a 1 676 m thick section of the Balikpapan Formation, at depths of between 75 and 1 737 m, with individual sands usually less than 30 m thick. Over 254 million barrels of oil have been produced since the field came on-stream, but with production declining rapidly the field was contracted out to the Tesoro Indonesia Company in 1973 for rehabilitation. The company succeeded in raising production levels to an average of over 9,000 barrels/day in 1976, but the level has since declined to average 4,257 barrels/day in April 1980 (Jefferies, 1980).

Modern seismic surveys which focused on the Mahakam delta and its vicinity, onshore and offshore, have revealed the existence of promising prospects in this part of the Kutei basin. Subsequent exploratory drilling led to the discovery of four fields: Attaka in 1970, Badak and Bekapai in 1972, and Handil in 1973.

Of these Attaka has proved to be the second most prolific. The field is 20 km offshore from the coast, north-east of the Mahakam River delta, in 60 m of water. It was the first commercial oil field to be discovered offshore in Kalimantan, and is the second largest offshore field in Indonesia. As the Attaka structure straddles the boundary between contract areas held by Union Oil and JAPEX Indonesia, a new operating unit was formed to explore and develop the field on a joint basis. Each company contributed equally to the 290 sq. km Attaka Unit, and all exploration, development and production costs are also shared equally.

The field was discovered in August 1970. The Attaka structure is a domal anticline complicated by transverse faults. It contains Early Tertiary to Quaternary sediments, the oil being in numerous deltaic sands deposited during a Late Miocene advance of the ancestral Mahakam River delta. The reservoir sands are from 1.5 to 30 m thick, and lie at depths of between 180 to 2 400 m. The oil is produced from thirty-four sandstone reservoirs lying at depths of between 600 and 2 400 m. Gravities of the oil range from 35 ° to 42 ° API, and sulphur content averages 0.1 per cent by weight (Schwartz, *et al.*, 1973). Initial oil production began in November 1972, only twenty-seven months after the discovery of the field. By November 1973 the field was producing 100,000 barrels of crude a day from fifty-two wells drilled from six platforms. By mid-1979, 228 million barrels of oil had been recovered from Attaka. As Fig. 5.4 shows, Attaka was the third largest oil field in Indonesia in mid-1979, responsible for 6.3 per cent of the total Indonesian output. Natural gas produced along with the crude oil is piped to an onshore liquid extraction plant at Santan, where it is processed into propane and butane for export.

The two other fields of Badak and Bekapai have also since become major fields, in spite of their being recent discoveries. The Badak oil and gas field operated by Roy M. Huffington, Inc. under a production-sharing contract, is located on the coast about 35 km north-east of Samarinda. It was discovered in January 1972 after three years of geological field-work and seismic and aeromagnetic surveys, followed by exploratory drilling in November 1971 in which gas, condensate and oil were encountered in numerous Middle Miocene to Pliocene sandstone reservoirs depths of between 1 370 and 3 350 m (4,500 to 11,000 ft.). The Badak structure is a gently folded, symmetrical anticline. Pay sands show excellent reservoir characteristics, with porosities ranging from 10 to 35 per cent. The oil from this field has a density of 38−39 ° API and a sulphur content of less than 0.1 per cent. Recoverable reserves are estimated to be more than 55 million barrels of hydrocarbon liquids and 7 trillion cu. ft of non-associated gas (Gwinn, *et al.*, 1974). Production of oil was 21,000 barrels per day and cumulative production 26 million barrels in mid-1979. LNG production at Badak was 3.7 million tons in 1980.

The Bekapai field is in the Mahakam Contract Area jointly held by

Total Indonesia and JAPEX Indonesia under a production-sharing contract with PERTAMINA. It is located offshore, about 15 km off the Mahakam delta. The Bekapai structure is a large faulted anticline, with commercial oil confined to the western fault block over an area of only 8 sq. km. The field was discovered in April 1972 after exploratory and delineation wells drilled in 35 m of water had confirmed the existence of an economic accumulation. The hydrocarbon-bearing sands are located at depths of between 1 300 and 1 600 m. They have excellent reservoir characteristics, with porosity varying from 25 to 35 per cent. Some of the reservoirs are prolific and can produce between 4,000 to 5,000 barrels of oil per day. Other reservoirs of lower yields have been located at greater depths (2 000 to 2 500 m). Production began in July 1974 from a single well yielding 5,000 barrels per day. With the completion of a multi-well platform production increased to 55,718 barrels per day for the first half of 1978, but declined to 47,649 barrels in mid-1979. By mid-1979 it had produced 69 million barrels of oil. The crude has a gravity of 40 ° API and a sulphur content of 0.08 per cent by weight (Matharel, *et al.*, 1976).

The biggest single offshore field discovered in Indonesia is Handil, with a potential of over 150,000 barrels per day. The Handil structure is located on the main channel of the Mahakam River. The swampy nature of the ground in the Handil area necessitates the use of offshore techniques for drilling. Modern seismic surveys were carried out in 1973, and the first well drilled in March/April 1974. Further drilling of delineation wells has confirmed the existence of a considerable thickness of hydrocarbon-bearing sands at depths of between 1 400 and 2 300 m. Some 150 reservoirs of oil and gas have been located within the 40 sq. km complex of Handil. Production began in June 1975 from four wells yielding 25,000 barrels per day, with output increasing to an average of 172,000 barrels per day by mid-1978, but declining to 166,912 in mid-1979. Cumulative production within this short period was nearly 171 million barrels. Handil is today the second biggest oil field in Indonesia, responsible for 10.4 per cent of the country's total output. The oil has a gravity of 31 ° to 36 ° API, and a sulphur content of 0.06 per cent (Magnier & Samsu, 1975; *Pertamina Today*, 1979; Sauphanor & Seguin, 1980).

OIL FIELDS IN THE TARAKAN BASIN

The Tarakan basin takes its name from Tarakan Island, off the coast of North-East Kalimantan. It is a 52 000 sq. km basin in which only two important oil fields have been discovered since the beginning of this century. Both these fields are located on small islands (Tarakan Island and Bunyu Island) formed by large anticlinal folds. The larger of the two fields is Pamusian discovered in 1905 on Tarakan Island. Cumulative production to mid-1979 was 204 million barrels. The producing formation, of Pliocene age, is geologically one of the youngest in the world to produce oil. It is poorly consolidated, and the fine sand and large quantities of water associated with the oil pose special problems. Almost all the cumulative production is from zones above 550 m, but since 1972 commercial oil has been found below 900 m. In 1971 the Tesoro Company was brought in by PERTAMINA to provide technical assistance to rehabilitate the Tarakan group of fields, which had reached the stripper stage of production. As a result of the operation, output increased from 0.7 million barrels in 1972 to 1.3 million barrels in 1973. The average daily production in the first half of 1979 was 2,495 barrels. The ultimate oil recovery from the Pamusian field is expected to be increased by about 15 per cent by this programme (Rowley, 1973). The oil produced from this field is a heavy, asphalt-base oil of 18 ° API gravity with 0.24 per cent sulphur by weight.

The smaller field in the island of Bunyu was discovered in 1923, and has recorded a cumulative output of 68 million barrels of paraffinic, 33 ° API gravity, low sulphur (0.06 per cent) oil by mid-1979. Pay zones are at depths of between 600 and 2 400 m, but the most recent wells drilled in the rehabilitation programme since 1974 are located at depths of between 3 000 and 4 000 m. Since 1974 production had increased from 4,000 barrels per day to 8,000 barrels per day (Soediono & Loucks, 1976). Production in mid-1979 was 7,980 barrels per day.

Exploration onshore by the ARCO Group resulted in the Sembakung discovery in 1976. Production from this field started in 1977. By mid-1979 the daily average production was 5,753 barrels, and the field had yielded a cumulative total of 5.5 million barrels of oil.

OIL FIELDS IN THE SALAWATI BASIN

Oil was first discovered in this basin by the Nederlandsche Nieuw Guinee Petroleum Maatschappij (N.N.G.P.M) when it drilled a discovery well at Klamono in 1936. By mid-1979 a total of 31 million barrels of reef oil had been recovered from the field, but daily production had declined to only 1,200 barrels.

The Salawati basin, which takes its name from an island off Irian Jaya, occupies the north-westernmost tip of Irian Jaya in the Vogelkop (Kepala Burung) peninsula, extending offshore to the west and south to the island from which the basin takes its name. It is a well-developed structural-stratigraphic basin with more than 4 600 m of marine Tertiary sedimentary rocks. The oil accumulations are entrapped within local carbonate reef culminations of Middle to Upper Miocene age superimposed on a larger reef - carbonate bank complex (Vincelette, 1973). These carbonate build-ups containing reefal organisms have a vertical relief of up to 490 m above the platform floor and possess exceptional porosities of 20 to 30 per cent. Production rates from individual wells are exceptionally high—from 20,000 to 32,000 barrels of oil per day.

N.N.G.P.M ceased exploration activities in the Salawati basin in 1960. In 1970 Petromer Trend signed a production-sharing contract with PERTAMINA covering an area of 4 500 sq. km in the basin. Seismic exploration of the area was started in April 1971, followed by exploratory drilling in September 1972 in three prospective areas: Kasim, Seget and Klanal. Seget and Klanal proved to be dry holes, but drilling in Kasim led to the discovery of a major oil field. Subsequent exploration led to the discovery of three other fields in the basin: Jaya, Walio and Kasim Utara. The Kasim and Jaya fields, although separate fields, are located on the same elongated reef mass. Both were originally part of a larger oil accumulation, but structural changes have resulted in the two separate accumulations present today. High porosities and excellent permeabilities have enabled individual wells in these two fields initially to produce at rates of 23,600 barrels per day (Kasim field) and over 25,000 barrels per day (Jaya field). The oil is a low sulphur (0.5 per cent), high API gravity (38 ° to 43 °) crude with an unusually low pour-point (−15 °F). Daily average production has since declined to 9,551 barrels per day (Kasim field) and 6,620 barrels per day (Jaya field) in mid-1979.

The Kasim Utara field is separated from the Kasim field by a 400 m channel. The reservoir consists of an upper porous reefal limestone 15 to 43 m thick underlain by a 90 m thick, highly fractured dolomite. The high permeability of the fractured zones has resulted in initial yields of 32,000 barrels per day from a single well (Vincelette & Soeparjadi, 1976). A new field, Kasim Barat, was discovered in 1975.

The Walio reef complex is part of the barrier and pinnacle reef system at the edge of a carbonate platform which extends south-westward from Klamono to Migool Island. The field appears to be a giant-sized one, covering an area of about 2 000 ha. Individual wells have penetrated more than 170 m of oil column, but the maximum production rate of 12,300 barrels per well was lower than the other fields in the basin (Redmond & Koesoemadinata, 1976; Vincelette & Soeparjadi, 1976). Production from the Walio field increased to a peak of 46,553 barrels per day in mid-1978, but dropped slightly to 43,772 barrels per day in mid-1979. Cumulative production from this major field was nearly 66 million barrels up to the first half of 1979.

OIL FIELDS IN THE NATUNA BASIN

Exploration for hydrocarbon accumulations in the Natuna basin started in the late 1968 when a consortium of oil companies headed by CONOCO signed a production-sharing agreement with PERTAMINA covering an area of 106 708 sq. km in Block B of the Natuna Sea. Gradual relinquishment of the original contract area reduced the size of the concession to 31 116 sq. km in 1979. Oil was discovered in 1974 in an elongated anticlinal closure trending NE–SW. The main axis of the closure is about 12.5 km long, with a width of 2.5 km. After confirmation wells were drilled, the 'A' platform in the field began production in January 1979, and by September 1979 it was producing 35,000 barrels of crude oil a day. Named Udang, the field is located about 1 000 km north of Jakarta, in the West Natuna basin. Output will be increased by another 20,000 barrels per day by 1981 when Udang 'B' goes on-stream (Mattes, 1979; *Petroleum News*, January 1980).

VI

Natural Gas

THE term 'natural gas' is used to describe the mixture of hydrocarbon gases, the principal component of which is methane, produced from underground reservoirs or from the processing of crude oil in refineries. Natural gas can either be 'associated gas' or 'non-associated' gas. Associated gas is that in direct contact with oil in the underground reservoir, much of the gas being actually dissolved in the oil. Associated gas can also occur in the form of a free gas cap overlying the oil in the reservoir. In both instances the gas supplies the energy necessary to liberate the oil from the reservoir rock and to lift it to the surface. A certain quantity of associated gas, either gas dissolved in the oil or gas from the free gas cap, is invariably produced with the oil. Each barrel of oil lifted to the surface and passed through surface separators will release from several hundred to several thousand cu. ft of associated gas originally dissolved in the oil. Such gas can be used for reservoir pressure maintenance or for industrial and commercial purposes.

Non-associated or 'dry' gas is the term applied to accumulations of natural gas which do not contain liquid oil in significant quantities and which can be produced independently to meet the market demand for gas. Several such gas fields have been discovered in Indonesia, at least two of which (Arun and Badak) are of giant size and have, since 1977, been developed for the large-scale production of liquefied natural gas.

The production of associated natural gas has increased in proportion to the increase in oil production, while that of non-associated natural gas has only become of significance since the development of the Arun and Badak gas fields. Figure 6.1 shows the production of

INDEX: 1971 = 100

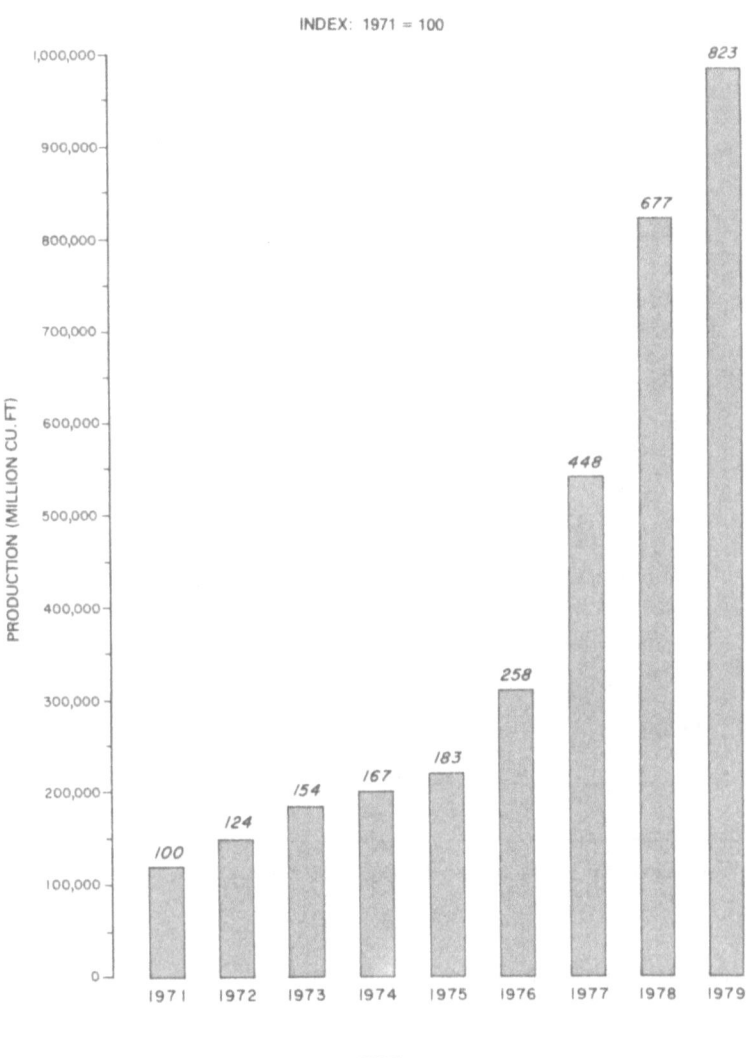

Fig. 6.1 Natural gas production in Indonesia, 1971–1979
Source: Direktorat Jendral Minyak dan Gas Bumi, Jakarta.

TABLE 6.1
NATURAL GAS PRODUCTION IN
INDONESIA BY TYPES OF CONTRACT, 1979

Company	Production (Million Cu. Ft)	Per Cent
State		12
PERTAMINA	122,479	
Lemigas	347	
Contract of Work		8
Caltex	28,636	
Stanvac	49,404	
Calasiatic & Topco	15	
Production-Sharing		80
Mobil Oil	370,826	
Asamera	7,114	
IIAPCO	24,204	
ARCO	63,118	
Calasiatic & Topco	150	
Union Oil	28,894	
Total Indonesie	42,568	
Inpex	64,632	
Roy M. Huffington Inc.	187,461	
Tesoro	2,030	
Petromer Trend	1,020	
Phillips	369	
CONOCO	5,190	
	998,547	100

Source: Direktorat Jendral Minyak dan Gas Bumi, *Petroleum and Natural Gas Industry of Indonesia*, December 1979 (Jakarta).

both associated and non-associated gas for the years 1971–9 while Table 6.1 shows the 1979 production by types of contract. Figure 6.2 shows production by company in 1979. The dominance of Mobil Oil and Roy M. Huffington Inc. is due to the large production of non-associated gas from their gas fields for LNG.

Historically, natural gas has been less valuable and less easily transportable than oil. The exploitation and sale of natural gas, whether associated or non-associated, have depended on the avail-

TOTAL PRODUCTION 998,457 MILLION CU. FT

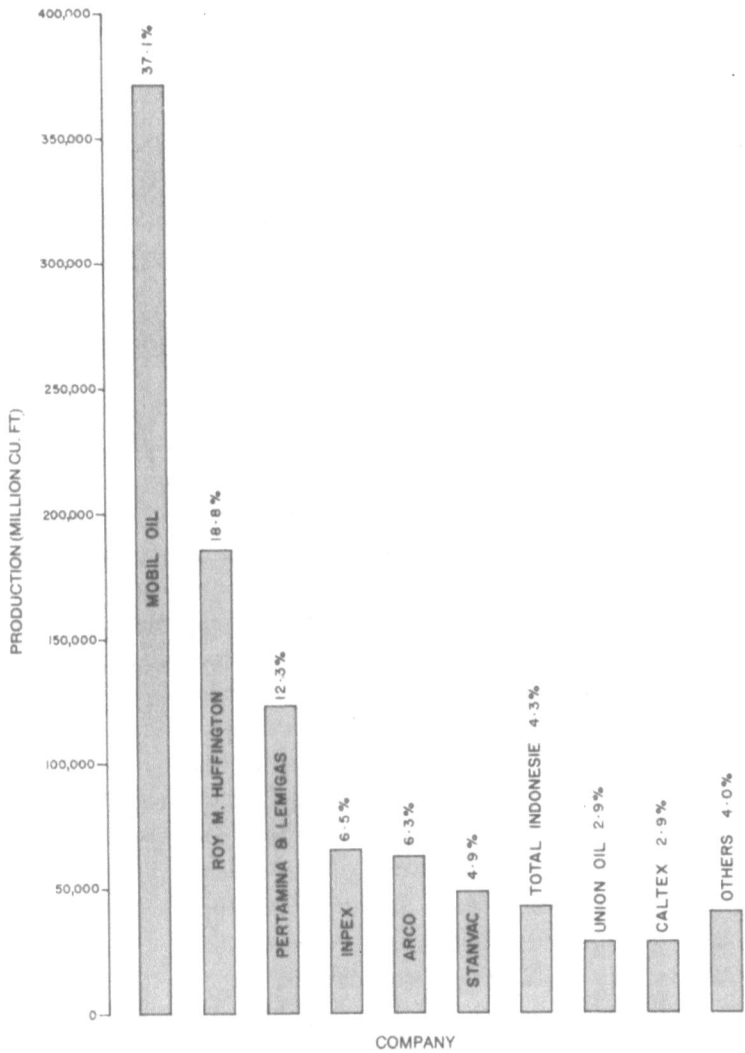

Fig. 6.2   Natural gas production, by company, in Indonesia, 1979

Source: Direktorat Jendral Minyak dan Gas Bumi, *Petroleum and Natural Gas Industry of Indonesia*, December 1979 (Jakarta).

ability of a market within a technically and economically feasible distance. There is, however, a major difference in the situations between associated and non-associated gas in that associated gas cannot normally be stored but has to be used as it issues from the oil wells, whereas in the case of gas fields producing non-associated gas the wells can be sealed and the gas stored until needed.

The volume of associated gas produced as a by-product of oil from oilfields is considerable. The problem of its disposal is compounded by the fact that most of this gas is available only at low pressure. In the absence of an economic market the gas is usually burnt or flared as waste. Even in the present-day context of high oil prices gas is being flared in the OPEC member countries at an annual rate that is more than sufficient to meet the entire energy needs of the developing economies of Asia (Symonds, 1978).

The alternatives to flaring the gas are to use or sell it directly as fuel, to use or sell it as a raw material for the manufacture of petro-chemicals and synthetic petroleum products, to compress and inject it into oil reservoirs as a means of enhancing oil recovery, or to liquefy the heavier components and market it as liquefied petroleum gas (LPG). In all cases additional investments are needed. However, depending on individual circumstances, the cost may be so great as to make it uneconomic to implement these measures, so that there is no practical alternative to flaring the gas as waste.

For much of the history of the petroleum industry in Indonesia the gas that was associated with oil was flared as waste. A small percentage was used by the oil industry as fuel in certain oil fields and refineries, and a further small percentage was compressed and re-injected into oil reservoirs for pressure maintenance. The Talang-Akar—Pendopo field of South Sumatra, for example, employed gas injection for pressure maintenance as early as 1927.

It was only in 1963 that Indonesia began to use natural gas for petro-chemical purposes, as feedstock for the urea fertilizer plant in Palembang. The plant, with a daily output of 300 metric tons, used natural gas from the Radja field in South Sumatra. The volume of gas involved was a mere fraction of the total volume of gas produced but this urea plant marked the beginning of a programme to make more effective use of a scarce resource. The need to minimize wastage and to diversify and increase the utilization of natural gas was given official recognition in 1968 when the National Natural Gas Utilization Committee was formed by governmental decree.

Even up to 1976, 60 per cent or more of the gas was either flared or otherwise lost, but with the increasing use of gas for pressure maintenance in oil production, for fuel, as feedstock in the chemical industry, and for power generation the percentage has dropped progressively. The associated gas produced in the more remote oil fields of Indonesia is too distant from existing markets to make it economically feasible to compress and transport it to such markets, especially when there are cheaper and more abundant supplies closer at hand. For this reason, the percentage of such gas put to productive use has not gone up in any dramatic way.

Nevertheless the Government of Indonesia, concerned with the mounting local consumption of oil (at present increasing at an annual rate of over 10 per cent) and very much aware of the fact that wherever natural gas can be used domestically in place of oil, such use would free the oil for export and increase foreign exchange earnings, is taking vigorous steps to ensure that its gas resources are utilized more efficiently. The gas conservation and utilization projects completed, under construction or under consideration are described below.

FERTILIZER PLANTS

Natural gas is an excellent raw material for the production of the derivatives of synthetic gas, notably ammonia, methanol and their end products. A fertilizer plant may produce ammonia or one of its derivatives, or it may be a fertilizer complex plant in which part of the ammonia produced is converted into other types of nitrogenous fertilizers, including urea. In a predominantly agricultural country such as Indonesia where large quantities of fertilizers are required and where abundant natural gas is available, the large-scale production of fertilizers from natural gas feedstock will meet two ends: free Indonesia from the need to import fertilizers at high cost, and minimize the wastage as well as maximize the use of its gas resources. Table 6.2 shows the fertilizer plant capacities using natural gas as feedstock projected to the year 1981. Of these plants, PUSRI I to IV began

TABLE 6.2

PROJECTED FERTILIZER PLANT CAPACITIES
USING NATURAL GAS AS FEEDSTOCK, 1981

Plant	Location	Capacity (Tonnes)
PUSRI I	Palembang, Sumatra	100,000
PUSRI II	Palembang, Sumatra	380,000
PUSRI III	Palembang, Sumatra	570,000
PUSRI IV	Palembang, Sumatra	570,000
PUSRI V	Palembang, Sumatra	300,000
Kujang	Cikampek, West Java	570,000
Kaltim	East Kalimantan	735,000
ASEAN Aceh	Aceh, North Sumatra	570,000
Total		3,795,000

Source: Directorate-General of Chemical Industries, BAPPENAS, PUSRI.
Note: The Table above does not include the Petro Kimie plants at Gresik with a total capacity of 665,000 tonnes as these plants do not use natural gas as feedstock.

operations in 1963, 1974, 1976 and 1977 respectively. PUSRI V is scheduled to begin production in 1981. The construction of the Kujang plant began in 1976 with commercial operations scheduled for 1979. It will produce both ammonia and urea using natural gas from the nearby offshore Arjuna field as feedstock. The Kaltim plant was originally designed to be the world's first floating fertilizer plant using two converted mineral ore barges anchored at Muara Badak off the east coast of Kalimantan to produce ammonia and urea. However the floating plant concept was abandoned in 1976 in favour of an onshore location at Bontang near the Huffco/PERTAMINA LNG plant. Commercial operation is projected to begin in 1981. The Japanese funded ASEAN Aceh plant was advanced as an ASEAN regional project with the Association of Southeast Asian Nations forming a holding company to acquire proprietary rights over the plant and be responsible for loan repayments. Under the scheme Indonesia will take over 60 per cent of the equity and the other ASEAN partners 10 per cent each. The plant will use natural gas from the Arun gas field as feedstock to produce urea and ammonia.

The projected capacities of 3 795 000 tonnes from all these natural gas-based fertilizer plants represent a very substantial jump when compared with the 8,000 tons per annum of fertilizers produced in Indonesia in the late 1960s *(Petroleum Economist,* July 1975, p. 248).

GAS PROCESSING PLANTS

Associated gas is also being used in gas separation and gas purification plants. A natural gas liquids (NGL) recovery plant is operated by ARCO in its offshore Arjuna field in the Java Sea. The plant has a through-put capacity of 230 million cu. ft per day of processed propane and heavier hydrocarbons. The output is about 14,000 barrels of liquefied petroleum gas (LPG) per day.

An LPG recovery plant to produce 4,480 barrels of LPG per day is under construction in Jatibarang, West Java. The output will be used domestically, freeing about 3.5 million barrels of crude oil for export per annum.

The Union Oil company has constructed a gas processing and crude stabilizing facility at Santan in East Kalimantan which will use natural gas from the offshore Attaka field for separation and processing into about 10,000 barrels of ethane and propane.

PIPED GAS FOR MULTIPLE USE

One of the most important steps being taken to increase natural gas utilization in Indonesia is the construction of an extensive gas pipeline network in West Java. Inaugurated in December 1978 the West Java Gas Transmission System when completed will supply 237 million cu. ft of gas per day to Jakarta for domestic consumption, to the giant Krakatau steel plant in Cilegon, to the Kujang fertilizer plant in Cikampek, and to cement factories in Cibinong. The main line of the network stretches from the Jatibarang oil field through Jakarta to Cilegon, a total distance of 334 km. One branch connects it to the Cibinong's cement factories and another to the Cikampek's fertilizer plant. The main supplies of gas will be from Jatibarang. Offshore supplies will be from the ARCO-operated Arjuna field and the Cemara-Parigi gas fields in the Java Sea, both connected to the network by underwater pipelines.

Other projects include PERTAMINA's plans to pipe some of the gas from its North Sumatra's fields to the city of Medan and the company's plans to pipe associated gas through an underwater pipeline from the Cities Service-operated Poleng oil field onshore to Surabaya for city gas, to fuel a power plant and to Gresik as feedstock for the fertilizer plant there.

OTHER PETROCHEMICAL PROJECTS

In line with the policies followed by other oil producers, Indonesia intends to divert an increasing proportion of its oil and gas for use as feedstock for the production of oil products and petro-chemicals. Among the petro-chemical projects based on natural gas under consideration are:

1. an olefins centre project in Aceh, which will use ethane feedstock extracted from natural gas from the Arun gas field. The integrated complex would be made up of:

> an ethane extraction unit of 450,000 ton/year capacity
>
> an ethane cracking plant of 350,000 ton/year capacity
>
> a polyethylene production unit of 150,000 ton/year capacity
>
> a vinyl chloride nomomer production unit of 110,000 ton/year capacity
>
> and a caustic/chlorine electrolytic unit of 70,000 ton/year of chlorine capacity.

The estimated cost of the total project is $1.6 billion (1980).

2. A methanol project in Bunyu Island in North-East Kalimantan which will use natural gas to produce 1,000 tons of methanol per day for the domestic and export markets. The estimated cost of $230 million is likely to be raised from a consortium of foreign companies.

LIQUEFIED NATURAL GAS

Exploration work in the 1970s by oil companies revealed the existence of two major gas fields in Indonesia: the Mobil/ PERTAMINA Arun field in North Sumatra, and the Huffco/ PERTAMINA Badak field in East Kalimantan. The Arun field, with in-place reserves of 17 trillion (million million) standard cu. ft, is one of the largest gas fields in the world; the Badak field is smaller, with in-place reserves of 7 trillion standard cu. ft.

The Arun field (Fig. 6.3) was discovered by Mobil in late 1971 in its

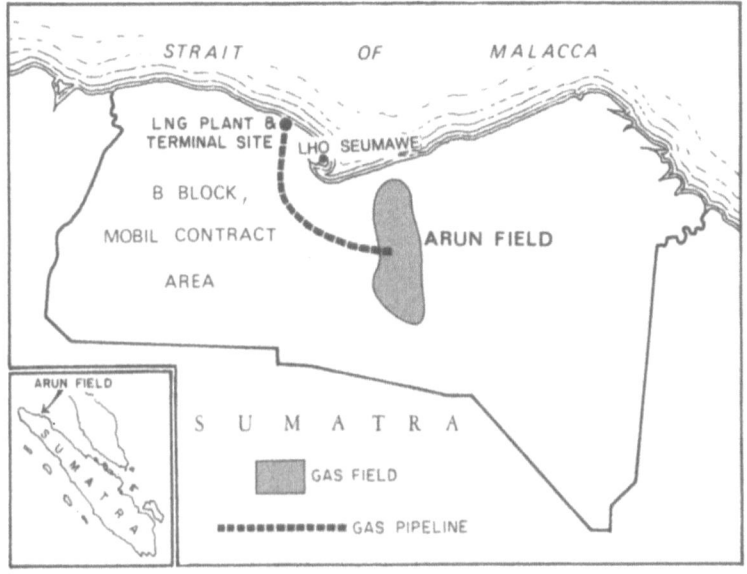

Fig. 6.3 The Arun gas field

B Block contract area on the north coast of Aceh Province. The field is located about 19 km south-east of the port of Lho Seumawe and about 241 km north-west of Medan. The Arun structure is a north-south trending feature about 18.5 km long by 5 km wide, with structural closure in all directions. Topographic elevations increase from a few m above sea-level at the north end of the field to about 30 m at the south end, at the foothills of the Barisan Mountains. The field is in a predominantly rice-growing area, and most of its surface land area is covered by padi fields.

The gas reservoir is a carbonate of Early and Middle Miocene age, underlain by a clastic sequence resting on basement rock of pre-Tertiary age. The reservoir is effectively enclosed in shales which, together with the organic material in the reef, are probably also the sources of hydrocarbons. The reservoir covers an area of about 17 000 ha. The average depth of the formation is about 3 048 m while the net pay thickness varies from 102 to 316 m. Gas produced from the reservoir has a methane content of 72 per cent, a carbon dioxide content of nearly 15 per cent, a small nitrogen content and a trace of hydrogen sulphide. The high potential of the reservoir is indicated by well rates of over a billion cu. ft per day, with significant quantities of condensate. The reservoir has a uniquely high reservoir temperature of 352 °F and high reservoir pressure of 7,100 psi. Wet gas-in-place has been calculated at 17.15 trillion standard cu. ft, or the equivalent of 3 billion barrels of oil. This will be sufficient to produce 2 billion cu. ft per day of gas for a twenty-year period (Graves & Weegar, 1973; Bramono, 1974; Alford, *et al.* 1975; Siagan & Stone 1976; Kingston, 1978).

The Badak field (Fig. 6.4) was discovered in 1972 by the Huffco Group working under a production-sharing contract signed with PERTAMINA in 1968. The field is located on the East Kalimantan coast, in a jungle-covered area about 120 km north-east of Balikpapan. Thirty-seven delineation and development wells drilled between 1972 and mid-1977 established Badak as a major gas field with reserves of

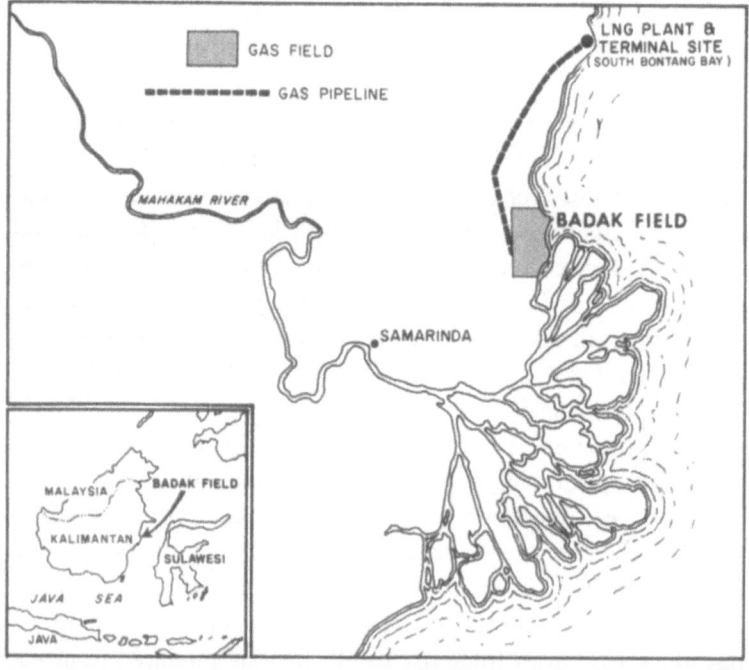

Fig. 6.4   The Badak gas field

more than 7 trillion standard cu. ft, large enough to support a commercial gas liquefaction plant.[1] The field spreads under an anticline approximately 11 km long and 5 km wide, within which over 70 separate associated and non-associated gas reservoirs have been proved at depths ranging from 1 158 to 3 353 m. Compared with the Arun field, the multiple-reservoir Badak field is very complex, with wide variations in reservoir size and well deliverability. The Huffco Group has also discovered additional new gas reserves in fields adjoining the Badak area. These reserves will be sufficient to provide additional capacity for the LNG plant.

Apart from these two gas fields, the Indonesian government has also announced the discovery of deposits of natural gas in two areas in South Sulawesi. The discoveries were made by British Petroleum Development of Indonesia Ltd. which holds exploration rights to a 23 309 sq. km area in South Sulawesi under production-sharing contract with PERTAMINA. The extent of the deposits is currently being determined.

It is apparent that the fossil fuel endowment of Indonesia is not limited to its oil resources but now includes substantial resources of non-associated natural gas. The demand for such gas as a source of energy has intensified in many industrial countries, particularly in Japan, because it is the least polluting of the fossil fuels. The recovery rate of non-associated gas from gas fields is from 50 to 90 per cent, a rate which is far higher than that of oil from oil reservoirs. But in comparison to oil which is easily tansported to end-users, natural gas, if it has to be transported to distant markets has first to be liquefied, moved to the consumer as LNG (liquefied natural gas), and then regasified before final use. The entire process is a complex one, involving the collection of gas from a network of wells, purification and cooling of the gas to $-160\,°C$, piping and storage, transportation in cryogenic LNG vessels, and, in the importing countries, the regasification of the LNG. All these operations must come on-stream together, so that there is little flexibility in the development of an LNG project. Additionally, the investment costs are very high:

[1] An independent survey by Degolyer & MacNaughton has put the recoverable reserves at 8.21 trillion cu. ft (*Petroleum News*, January 1980, p. 27).

Symonds (1978) estimated that at least $2.5 billion will be required to finance an LNG project with a capacity of 1,000 million cu. ft a day, the percentage breakdown of the costs being 12 per cent for production and delivery to the liquefaction plant, 40 per cent for the liquefaction plant and terminal, 36 per cent for marine transportation, and 12 per cent for the receiving terminal and regasification plant. In view of the massive investment costs involved' it is common financing and regulatory practice to ensure that such LNG projects be supported by proved reserves of at least twenty times annual deliveries in order to complete profitably the payout of the investment.

Soon after it was confirmed that sufficient quantities of gas existed to support large-scale LNG plants in Arun and Badak, PERTAMINA decided that, as there was no local market for the gas, it should be exported to earn additional foreign exchange. In late 1973 PERTAMINA concluded two separate sales contracts with Japanese and U.S. buyers. The Japanese contract involved five utility companies as end-users. The Arun and Badak plants would deliver 1,041 million cu. ft/day (roughly equal to 200,000 barrels of crude oil per day) or approximately 7.5 million tons of LNG per annum for a period of twenty years starting in 1977. The base price was indexed on a 90/10 ratio to the price of Indonesian crude oil and an annual escalation factor (3 per cent) respectively. On this basis the Japanese purchase price for the LNG was $2.40 per million Btu in 1978.

The U.S. contract of 1973 originally called for the delivery of 546 million cu. ft/day of LNG to the Pacific Lighting Corporation of Los Angeles for a period of twenty years, starting in 1978. The price was fixed at 63¢ per million Btu. Subsequently modifications to the contract were made, of which the most important was a 1978 agreement on a base price of $1.25 per million Btu, half of which price was tied to the price of Indonesian crude oil and half to the U.S. wholesale price index. The delivery date was also changed to the early 1980s.

The development of both the Arun and Badak gas fields will require not only the usual gas field processing equipment and the LNG processing, storage and loading facilities but also support installations and infrastructure such as plants for power generation, water supplies and cooling water systems, roads and harbour installations, internal communications systems and complete towns for LNG plant personnel at both Arun and Badak. An airport will also be built at Arun.

The Badak plant is designed to operate with two liquefaction trains initially, with provision being made for an additional two trains. Each train will be larger than any operating elsewhere in the world. The Badak LNG project had a shorter project schedule than that of Arun, and engineering activity began on it in November 1973. Completion of the project was targeted for March 1977 with full LNG production capacity being earmarked for July 1977. The project was completed on schedule and the first shipment of LNG to Japan took place in August 1977, with subsequent shipments following at regular intervals.

Natural gas from the Badak field is piped to three satellite gathering stations feeding the central station where the natural gas liquids are extracted. The gas is then dried and pumped along a 914 mm pipeline to the liquefaction plant in South Bontang Bay, 56 km away. The plant has two trains, each with an inlet processing capacity of 265 million cu. ft/day and an LNG output of 1.65 million tons/years. Each train has one cryogenic exchanger. There are four 600,000-barrel LNG storage tanks at the plant. The bay has a good harbour which needs only localized dredging to permit 4.4 million cu. ft (125 000 cu. m) LNG tankers to berth at the terminals.

Construction of the larger Arun project began in June 1975. The project has three phases. Phase one involves the construction of gas field and condensate producing facilities. This phase was completed in 1977, and the first shipments of about 12,000 barrels/day took place in October 1977. The second phase involves the construction of three liquefaction trains, each designed to have an inlet processing capacity of 280 million cu. ft/day and an LNG output of 1.6 million tons/year. This phase was completed at the end of 1978. The LNG from these trains, together with that from the Badak plant, will go to Japanese utility end-users. Phase three involves the construction of three additional trains which will liquefy natural gas for the U.S. market.

The cost of the Badak project was listed at $700 million and that

TABLE 6.3

LNG PRODUCTION AND RECEIPTS FROM SALES

Year	Production (Million Tons)	Receipts ($ Million)
1977 (August to December)	0.6	80
1978	3.7	540
1979	6.3	1,200
1980 (estimated)	8.4	2,800

Source: PERTAMINA.

of Arun at $950 million. Japan has provided the bulk of the loan finance, initially with a PERTAMINA-JILCO[1] loan of $900 million signed in 1974, supplemented by a Japan-Indonesia government-to-government loan of $187 million, an Indonesian government loan of $180 million, and a European-syndicated loan of $50 million. Subsequently additional financing was required to cover construction cost overruns at Arun and Badak, and in 1976 Japan, after prolonged negotiations, agreed to provide a further $320 million.

The transport arrangements to ship the LNG to Japan were made in 1976 by PERTAMINA, Burmah Tankers and the Japanese end-users. Under these arrangements Burmah Tankers will charter seven LNG cryogenic vessels of 4.4 million cu. ft capacity to be built by General Dynamics of St. Louis. Energy Transportation Inc. of New York will operate the vessels.

At current planned operating capacity PERTAMINA has estimated that the Arun and Badak LNG plants together will generate at least $18.7 billion in gross sales revenue over the 1977–99 period, of which Indonesia's share of net foreign exchange earnings will be $7.789 billion.

Since August 1977 when the first delivery of LNG was made to Japan, the production of LNG and receipts from its sale have expanded considerably as both the plants at Badak and Arun came into full operation (Table 6.3).

---

[1]JILCO is the acronym for Japan–Indonesia Liquefied Natural Gas Company, a new joint subsidiary of the five utility companies purchasing the Indonesian LNG.

TABLE 6.4
PRODUCTION CAPACITIES OF LNG, EXISTING AND PLANNED, ARUN AND BADAK
(*Million Tons/Year*)

Capacity	Arun	Badak (Bontang)	Total
Current (1980)	4.7	3.7	8.4
Planned expansion	7.3	3.2	10.5
Total	12.0	6.9	18.9

Source: PERTAMINA.

It is apparent that earnings of such magnitude represent a major contribution to Indonesia's net revenues from its hydrocarbon resources. In keeping with its policy of maximizing revenues derived from these resources PERTAMINA is now seriously exploring the possibility of expanding its production of LNG for the Japanese and U.S. markets, through a two-train expansion of the Bontang plant at Badak and a similar two-train expansion of the Arun plant.[1] The output from these four new trains will go to Japanese end-users. In addition, the existing 1973 contract with a group of California utility buyers to supply LNG from Arun to California will, when finally ratified between the two parties, result in three-train expansion of the Arun plant. The total cost of the seven-train expansion will be in the region of $2.4 billion (Cowper, 1980).

The planned expansion, if realized, will result in a 125 per cent increase in the production capacities at Arun and Badak by 1984–5 (Table 6.4).

The former President-Director of PERTAMINA Major-General Piet Haryono in his speech at the World LNG Conference in Kyoto in April 1980 and in another speech in Jakarta in May 1980 forecasted that by the end of the decade Indonesia could have as many as eighteen LNG trains, and that the foreign exchange earnings from

---

[1] It was announced in November 1980 that an agreement had been reached between Indonesia and Japan for Indonesia to supply an additional 3 million metric tons of LNG annually to Japan. The Arun LNG plant's capacity will be increased by two trains to meet this demand (*Straits Times*, Singapore, 6 November 1980).

LNG could be as high as $6 billion annually. Such a large-scale expansion in output would be based on the planned increases at Arun and Badak as well as the development of a newly-discovered large gas field in the South China Sea near Natuna Island. The field is in an ESSO contract area, and an ESSO team has begun studies on the feasibility of setting up an LNG plant in Natuna Island (*Petroleum Economist*, May 1980).

PRESENT AND FUTURE TRENDS

The sum of the efforts described above is a significant increase in the percentage of natural gas used for productive purposes and a corresponding fall in the percentage flared or otherwise lost (Table 6.5).

The percentage used for various purposes in the oil fields has remained relatively constant. The main change in this period have been in the percentages of natural gas used for other productive purposes, notably for the production of LNG, accounting for 38 per cent of the total natural gas output in each of the years 1978 and 1979).

A more detailed breakdown of natural gas utilization is given in Table 6.6. Local sales include natural gas used for city gas, for the Kujang fertilizer plant, cement factories and the Krakatau steel plant.

TABLE 6.5
NATURAL GAS UTILIZATION IN INDONESIA, 1974–1979

	Percentage of Total Gas Produced					
	1974	1975	1976	1977	1978	1979
Used in oil fields for fuel, gas lift, pressure maintenance	25	26	22	20	27	26
Other uses	11	11	18	32	46	51
Flared or otherwise lost	64	63	60	48	27	23
Total	100	100	100	100	100	100

Source: Data from Direktorat Jendral Minyak dan Gas Bumi, *Petroleum and Natural Gas Industry of Indonesia, 1974–9* (Jakarta).

TABLE 6.6
NATURAL GAS UTILIZATION IN INDONESIA,
BY END USES, 1979

	Quantity (Million Cu. Ft)	Per Cent
As fuel gas, for gas lift and pressure maintenance in oil fields	262,112	26
Local sales	32,229	3
PUSRI	47,840	5
Refinery	8,238	1
LPG production	38,092	4
LNG production	383,303	38
Lost/flared	226,643	23
Total	998,457	100

Source: data from Direktorat Jendral Minyak dan Gas Bumi, *Petroleum and Natural Gas Industry of Indonesia* (Jakarta), December 1979.

Gas is also used in the PUSRI fertilizer plants, and in the production of liquefied petroleum gases (LPG). Altogether the percentage of such uses was only 13, not a large percentage increase from 1974 and 1975, although the quantity of natural gas (largely associated gas) used for these purposes has gone up in absolute terms. Table 6.6 shows the dominant position occupied by LNG production from non-associated gas. It should be noted that the inclusion of LNG in the Table has resulted in depressing the percentage lost through flaring, usually employed to dispose of unwanted associated gas produced in oil fields.

Indonesia's total gas reserves are estimated at 37 trillion cu. ft (*Petroleum Economist*, August 1980). In order to encourage foreign companies to invest in gas exploration and development Indonesia may increase domestic gas prices which at the moment range between 50¢ and $1.92 per million Btu, well below the equivalent in oil prices. Increased gas prices of $2.50 per million Btu would also save the government up to $400 million a year in fuel oil subsidies, while still making it cheaper to use than oil (*Indonesian Observer*, 7 August 1980).

The last quarter of the twentieth century may be regarded as a milestone in the history of the Indonesian petroleum industry in that, with the first shipment of LNG to Japan in 1977, non-associated natural gas from gas fields in Indonesia joined crude oil as an export commodity and a major foreign exchange earner. The contribution of natural gas, both associated and non-associated, to the economy is certain to increase in future years as new gas fields are discovered and developed, and greater use is made of natural gas as a domestic energy source, for petro-chemicals and synthetic petroleum products, as feedstock for fertilizer production as well as for reservoir pressure maintenance. LNG may indeed displace crude oil as Indonesia's largest foreign exchange earner in the next decade, particularly if oil exports decline as a result of declining production and increasing demand for domestic consumption.

VII

Reserves and Resources

Reserves

ASSESSMENTS of resources belong to the realm of educated guesses. Conceptually distinct from resources and occupying a higher ranking in regard to precision, certainty and accuracy in mineral resources terminology in the term 'reserves'. As defined in 1974 by the U.S. Bureau of Mines and the U.S. Geological Survey 'reserves' are that portion of an identified resource from which a usable mineral and energy commodity can be economically and legally extracted at the time of determination. Reserves are further subdivided into 'proved', 'probable' and 'possible'. 'Proved' or 'measured' reserves are material for which estimates of the quality and quantity have been computed, within a margin of error of 20 per cent, from analyses and measurements from closely spaced and geologically well-known sites. 'Probable' or 'indicated' reserves are material for which estimates of the quality and quantity have been computed partly from sample analyses and measurements and partly from reasonable geologic projections. 'Possible' or 'inferred' reserves are material in un-explored but identified deposits for which estimates of the quality and size are based on geologic evidence and projection. The conceptual organization of these items is shown in Figure 7.1.

Proved reserves represent expensive capital committed to inventories, and in the petroleum industry, as in any other business organization, the policy is not to tie up more capital than it needs to. Proved reserves have thus seldom exceeded 30 to 35 years of world-wide consumption, while in the United States, such reserves have seldom gone beyond 12 years of consumption (Safer, 1978). However, reported figures of proved reserves do not indicate the parameters

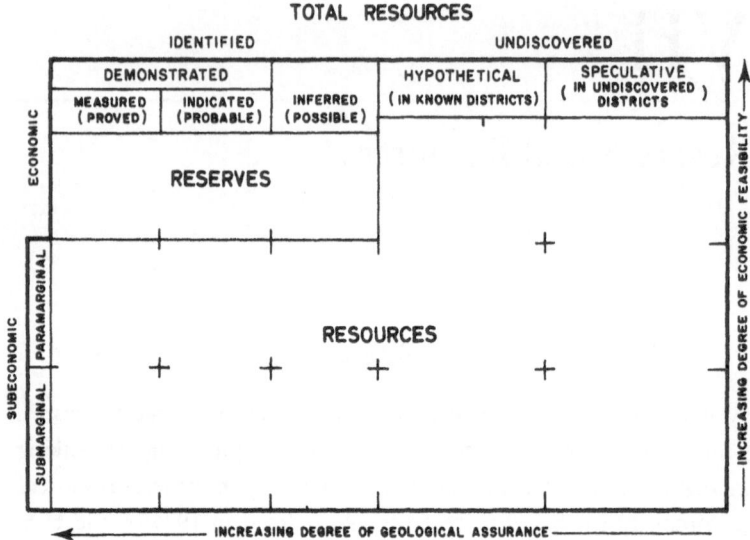

Fig. 7.1 Conceptual organization of the terms 'reserves' and 'resources'
Source: Based on U.S. Geol. Surv./U.S. Bur, of Mines, 1974.

whereby the operator measures these reserves, and there is no authoritative world body which collects and publishes proved reserves on a standardized basis. The most reliable estimate is that based on careful measurements of the physical characteristics of the reservoir rock in order to arrive at the amount of petroleum in place. Further measurement of the characteristics of the reservoir fluids and the production mechanics and the effect of stimulation can provide a good estimate of the recovery factor and hence the proved reserves calculated at that point in time under current economic conditions (Ion and Jamieson, 1966).

The world's estimated oil reserves are calculated on the basis of current prices as well as existing techniques, and each rise in oil prices will theoretically add to these reserves. Higher oil prices will also provide the economic incentive to improve recovery methodology. In the U.S., for example, reserve estimates are based on an average conventional recovery rate of 30–35 per cent of the oil-in-place. Enhanced recovery techniques using chemicals or heat to supplement the conventional secondary recovery techniques of water or gas injection may increase the recovery rate to 40–45 per cent. It is conceivable that future rates may increase to as high as 50–60 per cent of the oil-in-place (Meyer and Hocott, 1977). Such an advancement will greatly increase the reserves of currently known oil fields, more so if it is borne in mind that even a 1 per cent increase in average recovery from the world's oil fields would be equivalent to adding a year's production to the ultimate recovery likely to be obtained from them (Hobson and Tiratsoo, 1975). However, according to one major oil company, enhanced recovery techniques cannot be expected to contribute significantly to world oil supplies in the next decade or so (*Petroleum Economist*, May 1979).

While published compilations of national proved reserves are obviously useful, their reliability can vary from country to country. Published reserves estimates may, for commercial, political or military reasons, differ from estimates made by oil companies or by governmental bodies. Moreover, the volume of oil-in-place even in a well delineated field cannot be precisely determined. It is usual for an oil company to assess the volume of reserves in any field in which oil is discovered, and to adjust such assessment in the light of knowledge gained through the drilling of appraisal and production wells. As development of the field proceeds, proved reserves are generally increased as reserves originally designated as 'probable' or 'possible' are moved into the 'proved' category. Estimates of reserves are usually made on the basis of the production system currently employed and not on the basis of existing methods of technology. Thus the value of estimates of proved reserves lies not in providing an accurate picture of absolute availability but in providing (1) an indication of the location of known reserves and (2) an assessment of the quantity of oil available with present techniques and at present cost.

PROVED RESERVES OF SOUTH-EAST ASIA
AND INDONESIA

Table 7.1 shows the published estimated proved reserves of oil and gas in South-East Asia as at January 1980. The total proved reserves of 14,250 million barrels of oil for the region amounted to only 2.2 per cent of the world's reserves. The proved gas reserves constituted also 2.2 per cent of the world's gas reserves. Indonesia had two-thirds of

TABLE 7.1
ESTIMATED PROVED RESERVES OF OIL AND GAS IN SOUTH-EAST ASIAN COUNTRIES, JANUARY 1980

Country	Oil — Estimated Proved Reserves (Million Barrels)	Per Cent	Gas — Estimated Proved Reserves (Billion Cu. Ft)	Per Cent
Indonesia	9,600	67.4	24,000	42.2
Malaysia	2,800	19.6	17,000	30.0
Brunei	1,800	12.6	7,700	13.5
Philippines	25	0.2	—	—
Burma	25	0.2	135	0.2
Thailand	—	—	8,000	14.1
Total	14,250	100.0	56,835	100.0

Source: *Oil and Gas Journal*, 31 December 1979.

South-East Asia's proved oil reserves, and 42 per cent of its proved gas reserves.

The published estimated proved reserves of Indonesia[1] have fluctuated considerably between the years 1957 and 1980 (Table 7.2). In general, however, there has been an upward trend with time, though not of the order of magnitude of the published world proved reserves between the years 1945 and 1973 when these increased ten times—from 58.9 billion barrels to 600.3 billion barrels (Hobson and Tiratsoo, 1975). Nevertheless, if the average of the 1957 estimates is taken as a base, the estimated proved reserves of Indonesia have increased by 80 per cent between then and 1980. The estimates of 15 billion barrels in 1975 and 14 billion barrels in 1976 were scaled back to 10.5 billion barrels in 1977 upon re-evaluation of the country's existing fields by Indonesian experts. A similar re-evaluation of the gas fields resulted in an upward revision of the gas reserves from 15 trillion to 24 trillion cu. ft.

[1] The government of Indonesia does not publish any official estimates of the oil and gas reserves in the country.

TABLE 7.2
ESTIMATED PROVED RESERVES
IN INDONESIA, 1957–1980

Year	Proved Reserves of Oil (Billion Barrels)	Proved Reserves of Gas (Billion Cu. Ft)
1957	3.0 to 7.7	—
1960	9.5	—
1965	9.5	—
1970	10.0	—
1972	10.4	4,500
1974	10.5	15,500
1975	15.0	15,000
1976	14.0	15,000
1977	10.5	24,000
1978	10.0	24,000
1979	10.2	24,000
1980	9.6	24,000

Sources: U.N. ECAFE (1959); *Oil and Gas Journal*, various years.

EXPANDED PROVED RESERVES

Grossling (1976) has provided a statistical definition of expanded proved reserves which may also be applied to South-East Asia and Indonesia. Proved reserves (R_1) are defined as amounts of petroleum which are recoverable, with a 90+ per cent probability, from discovered fields within present economic and technological limits. Such amounts can be estimated within ±25 per cent. Additional amounts of petroleum may exist beyond the known parts of the field and as extensions of them. Such additional amounts may be inferred with varying degrees of probability, and together with the proved reserves, will constitute the expanded proved reserves (R_2).

He estimates that for most of the non-OPEC countries an additional quantity equal to the proved reserves can be inferred with a probability of 0.8, and a similar quantity can also be inferred with a probability of 0.5. The expected value of the expanded proved reserves would therefore be given by the equation $R_2 = (1 + 0.8 + 0.5) R_1 = 2.3R_1$

For OPEC countries the expected value of the expanded proved

TABLE 7.3
EXPANDED PROVED RESERVES
IN SOUTH-EAST ASIA, 1980

Country	Oil (Million Barrels)		Gas (Billion Cu. Ft)	
	Proved Reserves (R_1)	Expanded Proved Reserves (R_2)	Proved Reserves (R_1)	Expanded Proved Reserves (R_2)
Indonesia[1]	9,600	14,400	24,000	36,000
Malaysia	2,800	6,440	17,000	39,100
Brunei	1,800	4,140	7,700	17,710
Philippines	25	57.5	—	—
Burma	25	57.5	135	310
Thailand	—	—	8,000	18,400
Total	14,250	25,095	56,835	115,520

Source: As in Table 7.1.

[1] For Indonesia, an OPEC member, $R_2 = 1.5R_1$; for the other countries, $R_2 = 2.3R_1$.

reserves should be lowered to $R_2 = 1.5R_1$ for two main reasons. First, it is considered necessary for OPEC to maintain its leverage on proved reserves in order to enable it to increase crude oil prices. Hence there must be a tendency to exaggerate the proved reserves figures. Second, there would also be a tendency by individual OPEC members to inflate such figures in order to enhance their leverage within OPEC (Grossling, 1976).

Using this definition the expanded proved reserves of South-East Asia (including Indonesia) are shown in Table 7.3.

Following on the concept of expanded proved reserves, Grossling (1976) introduced the concept of expected value of recoverable discoveries (EVRD), defined as the sum of the cumulative production in a given country plus the expanded proved reserves. The EVRD of South-East Asian oil-producing countries is given in Table 7.4.

By dividing the expanded proved reserves by the EVRD and expressing the result as a percentage, it is possible to rank the oil-producing countries of South-East Asia by the percentage of the EVRD that remains to be produced (Table 7.5). The higher the

TABLE 7.4

THE EXPECTED VALUE OF RECOVERABLE DISCOVERIES
(EVRD) OF CRUDE OIL, SOUTH-EAST ASIA
(Million Barrels)

Country	Cumulative Production to mid-1979	Expanded Proved Reserves	EVRD
Indonesia	8,207	14,400	22,607.00
Malaysia	504	6,440	6,944
Brunei	1,492	4,140	5,632
Burma	411.5	57.5	469
Philippines	2	57.5	59.5
Thailand (mid-1978)	1	0.46	1.46
Total	10,617.5	25,095.46	35,712.96

Source: As in Table 7.1.

percentage the larger the quantity of oil that remains in the ground awaiting production.[1] This quantity is relative to the expanded proved reserves of individual countries. Such reserves, as Table 7.4 shows, vary considerably among the countries. The percentages shown in Table 7.5 are an indication of the stages of petroleum development of the various countries. The Philippines, for example, has prospective areas but have not yet begun to produce oil in significant quantities, so that its percentage of EVRD that remains to be produced stands at almost 97. In contrast, Burma with a long history of petroleum development has only 12 per cent of its EVRD left. Indonesia has also a long development history, but such development has been made on a much larger petroleum resource base, so that almost two-thirds of its EVRD still remains to be produced.

RESERVES PRODUCTION RATIOS

Models of the life-time of hydrocarbon deposits are employed by governments and other agencies for a variety of purposes, especially in planning for economic development. The figure most often quoted

[1] The percentages can vary directly with the exploration effort; low figures may also be due to inadequate exploration.

TABLE 7.5

RANKING OF SOUTH-EAST ASIAN OIL-PRODUCING COUNTRIES BY EVRD THAT REMAINS TO BE PRODUCED, MID-1979

Country	Percentage of EVRD that Remains to be Produced (Expanded Proved Reserves) EVRD	Ranking
Philippines	96.6	1
Malaysia	93.7	2
Brunei	73.5	3
Indonesia	63.7	4
Thailand	31.5	5
Burma	12.3	6

is the reserves/production ratio, whereby the life-time of the reserves is calculated by simply dividing the proved reserves by the year's production. The life-time of the reserves will be correspondingly shorter if an annual production increase is assumed.

The ratios between the world published proved reserves and annual production since the 1930s are as follows:

1930	18
1940	16
1950	21
1960	39
1970	36
1976	35

Sources: Hobson and Tiratsoo (1975); Kehrer (1978).

In spite of large production increases, the discovery of new reserves has allowed the reserves/production ratio of over thirty years established in the 1960s to be maintained in the 1970s. The ratio would be extended to a hundred years (from 1976) if the estimated ultimate recovery (EUR) were divided by the world oil production for 1976 (Kehrer, 1978).

Table 7.6 shows the reserves/production ratios for Indonesia for various years. The 1979 ratio is only 16.5 years; the ratio would

TABLE 7.6
RESERVES/PRODUCTION RATIOS, INDONESIA

Year	Production (Million Barrels)	Proved Reserves (R_1) (Million Barrels)	Expanded Proved Reserves ($R_2 = 1.5\,R_1$) (Million Barrels)	Proved Reserves/ Production Ratio (Years)	Expanded Proved Reserves/ Production Ratio (Years)
1965	177	9,500	14,250	54	80.5
1970	311	10,000	15,000	32	48
1975	477	15,000	22,500	31	47
1979	580	9,600	14,400	16.5	25

Sources: Data from U.N., ECAFE, 1959; *Oil and Gas Journal*, 31 December 1979; *Pertamina Bulletin*, February 1978; Direktorat Jendral Minyak dan Gas Bumi, *Petroleum and Natural Gas Industry of Indonesia* (Jakarta), December 1979.

increase to twenty-five years if the expanded proved reserves are used as the basis for calculation.

Table 7.6 shows that the reserves/production ratios have declined considerably since 1965. This is due to the fact that the almost 330 per cent increase in oil production recorded between 1965 and 1979 has not been matched by a corresponding increase in the amount of proved reserves, the published proved reserves having only increased by 1 per cent during that period of time.[1] In contrast, the increase in the world reserves of oil recorded over the same period of time was in the region of almost 80 per cent.

The proved reserves of gas in Indonesia have increased substantially, from 2.8 trillion cu. ft in 1970 to 24 trillion cu. ft at the end of 1979 (*Oil and Gas Journal*, 31 December 1979). This represents an increase of 850 per cent. With a production rate of 998 billion cu. ft in 1979 the life-time of the reserves at that production level would be twenty-four years. A different source has indicated that the proved reserves of natural gas was 37 trillion cu. ft in January 1979, and with

[1]As indicated earlier, the published proved reserves of 15 and 14 billion barrels in 1975 and 1976 were scaled back to 10.5 billion barrels in 1977 upon re-evaluation of the oil fields by Indonesian experts.

production at 998 billion cu. ft, the life-time of the reserves would therefore be thirty-seven years (*Petroleum Economist*, August 1979).

The reserves/production ratios are a simple approach to a complex problem in that neither the present oil reserves nor the present rate of production are static but are constantly changing over time. Reserves may be increased by new discoveries, by the development and application of new technology such as enhanced recovery techniques, deeper drilling and/or by new economic incentives (higher oil prices, improved infrastructural facilities in remote areas, fiscal incentives, etc.) which may convert previously uneconomic prospects into economic ones. Simultaneously reserves are being depleted as oil is extracted, the reserves declining if the rate of depletion is greater than the amount of new reserves added, and vice versa. Where there is a net increase the added reserves can physically support a higher level of production, the reverse holding true where there is a net decrease. Production itself is determined by a combination of geological, technological, economic and political factors, and does not remain constant.

In the calculation of these ratios it is assumed that production will be maintained at the existing level until the last barrel of oil is extracted. In actuality production will vary because the dynamic nature of an oil field works against the maintenance of the same level of production or even a continual increase until complete exhaustion (Kehrer, 1978). The production rate rises continuously until it cannot be supported by the reserves in the field, at which time it declines until the field becomes uneconomic to exploit.

The reserves/production ratios as models of the life-time of the oil reserves are therefore only useful as indicators of the order of magnitude of the ratios of the reserves to production. In viewing these figures one needs also to bear in mind the point that Safer (1978) has made, that when private oil companies project declining reserves more than five years out, they do so with a view to using these forecasts as the basis for budgeting funds for exploration. But when governments make such forecasts of declining reserves, they use these to draw doomsday conclusions about impending oil shortages. In the past such conclusions have repeatedly been proved to be premature.

Resources

Such estimates of the petroleum resources of Indonesia as are attempted from time to time can and do vary widely due partly to the use of differing definitions and terminology but more especially to the methodologies used to construct the future availability of these resources.

The term resource itself is an imprecise one. Various definitions exist. For example, Ion (1975) defines it as 'the total amount of the resource base which is estimated to be probably recoverable for the benefit of man. This estimate will be based on both knowledge and reasonable conjecture regarding location and probable recovery techniques.' Kehrer (1978) regards resources as identified plus undiscovered deposits, regardless of their present recoverability. The United Nations Committee on Natural Resources (1975), on the other hand, defines resources as 'deposits which are of no advantage at present but which may become more profitable in the future under more favourable economic or technical conditions'. In all cases the inclusion of unproven (hypothetical, speculative) deposits in estimations of resources adds a large degree of uncertainty to such estimations.

At present all published estimates of petroleum resources refer to primary production of petroleum from conventional oil or gas fields. The rates of recovery from such production are usually between 30–35 per cent of the petroleum-in-place. Given the existence of such petroleum-in-place the availability of petroleum is dependent on technological and economic factors.

In general the amount of technology that can be employed in the extraction of a mineral varies directly with the value of the mineral. As the price of crude oil has now increased from its former level of $3 per barrel to over $30 per barrel (1980) substantially more technology can be applied for its extraction. Also very large inputs of funds and manpower are being injected by many governments and large private oil companies into research on the technology of petroleum extraction from conventional sources as well as alternative sources such as tar sands and oil shales. Improvements in the technology of extraction will lead to an increase in the rates of recovery and hence

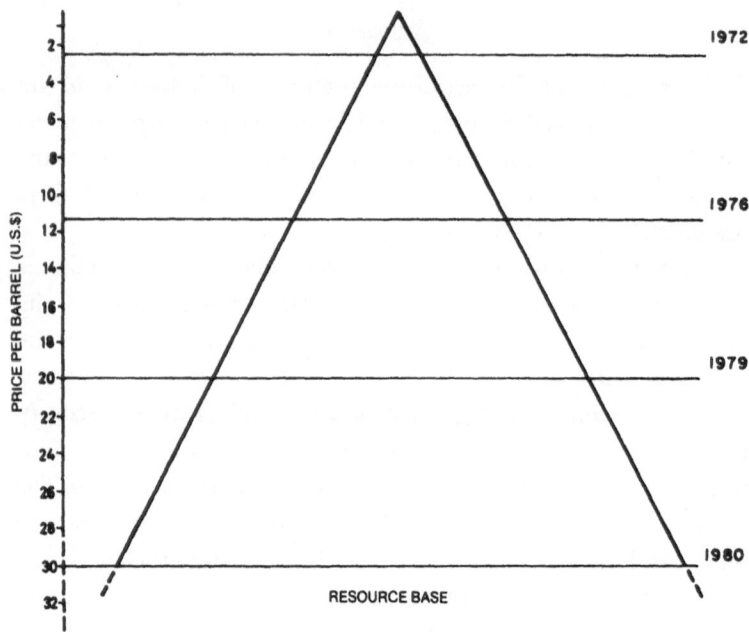

Fig. 7.2   Availability of oil and gas as a function of time and price
Source: Based on Barnea (1977), p. 27.

an expansion in the resource base. For example, with the price increases the secondary recovery of petroleum is now economically feasible in most onshore fields in the world.

The resource base will also expand with the multifold increases in petroleum prices because the resources that were previously uneconomic to extract have now become economic. The relationship between resource availability and price is illustrated in Figure 7.2. With further price increases in the future a position will be reached when the production of petroleum from coal or renewable resources will become economic (Barnea, 1977).

UNDISCOVERED HYDROCARBON RESOURCES
In order to estimate the undiscovered hydrocarbon (oil and gas) resources of Indonesia it is necessary to set out the various terms and parameters employed in the exercise. These and the estimate itself are

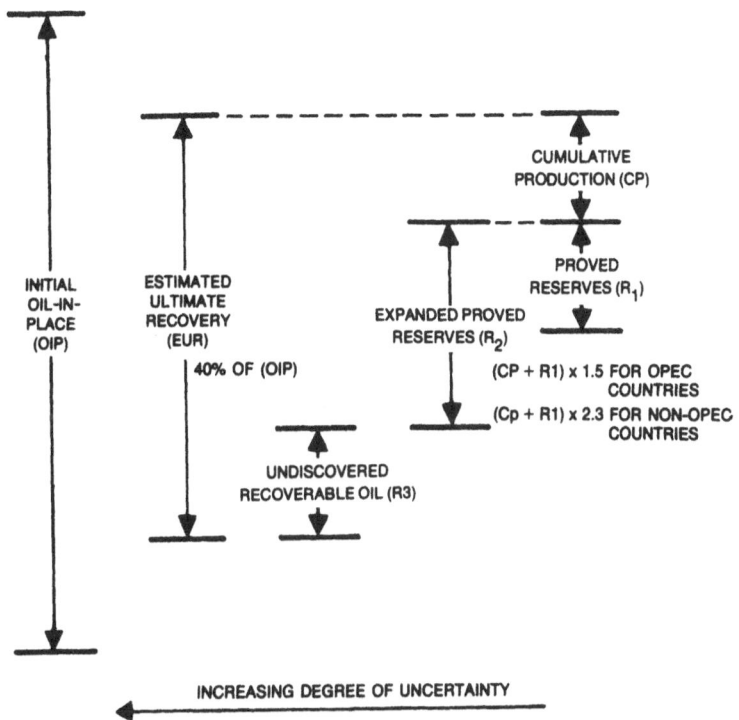

Fig. 7.3 Conceptual framework of the resource and reserve
terms for oil and gas

Source: Based on Grossling (1976) Fig. 1.

based on Grossling's *Window on Oil*, published by the *Financial Times*,
London, in 1976.

The conceptual framework of the terms employed is given in Figure
7.3. These terms are:

Oil-in-Place (OIP). This is the initial amount of oil in undiscovered
and discovered fields, before exploitation. The OIP is difficult to
estimate and is subject to constant revisions with increasing knowledge about the petroleum geology of an area.

Estimated Ultimate Recovery (EUR). This is the amount of oil
that can be recovered from the oil-in-place, under existing economic
and technological conditions. Both these conditions can and do
change with time, so that the EUR will also vary. Grossling has taken

40 per cent of the OIP as representing present-day average conditions of recovery.

Cumulative Past Production (CP). This is the total amount of oil produced from discovered fields.

Proved Reserves (R_1). This is defined as the amount of oil which can be recovered from known fields within existing economic and technological limits.

Expanded Proved Reserves (R_2). This is the expected amount of oil that can be statistically inferred to exist from revisions and expansions of known fields.

Undiscovered Recoverable Oil (R_3). This is the difference between the EUR and the cumulative past production and expanded proved reserves, as given in the equation $R_3 = EUR - (CP + R_2)$

In order to establish the amount of undiscovered recoverable oil resources in an area it is necessary to find out the estimated ultimate recovery (EUR) of that area. Grossling (1976) in his study of the petroleum prospects of the developing world took into account the following factors when estimating the EUR on a world-wide basis:

1. the size of the petroleum prospective areas,

2. the density of drilling for petroleum,

3. the outcome of past exploration and development,

4. the number of giant fields per million square miles of prospective area,

5. the expected value of recoverable discoveries per unit of prospective area, and

6. the bench-mark areas for estimating the EUR. Based on the above factors he arrived at a range of 70,000 to 200,000 barrels of oil and 400 to 1,000 million cu. ft of gas per sq. ml (2.59 km²) of prospective area as the EUR from continental size blocs. The equations are:

for oil EUR = (70 to 200) $\times 10^3 \times$ prospective area

for gas EUR = (400 to 1,000) $\times 10^6 \times$ prospective area

UNDISCOVERED HYDROCARBON RESOURCES
OF INDONESIA

It is possible, in applying Grossling's estimation procedure to Indonesia, to arrive at different estimates of its undiscovered hydrocarbon resources because of differences in the estimates of the country's prospective area.

Grossling in estimating the petroleum prospective areas of the world, consulted the following sources: tectonic and geologic maps and articles on petroleum prospects of various regions and countries; regional maps in annual issues on foreign petroleum developments of the 'Bulletin of the American Association of Petroleum Geologists, 1949–75'; and, for areas of the continental shelves down to the 200 m water depth, *International Boundary Study, Series A, Limits in the Seas*, No. 46, 21 August 1972, *The Geographer*, U.S. Department of State. He arrived at the following figures for Indonesia: onshore prospective areas, 390,000 sq. mls; offshore prospective areas, 855,000 sq. mls; total prospective areas, 1,245,000 sq. mls.

Ismet Akil and G. A. S. Nayoan (1973) of the Geological Department of PERTAMINA in their article 'Notes on offshore petroleum development in Indonesia' have estimated the prospective areas with Tertiary sedimentary cover in Indonesia as, onshore 1 700 000 km^2 (656,000 sq. mls); offshore (200 m water depth) 1 400 000 km^2 (541,000 sq. mls); total 3 100 000 km^2 (1,197,000 sq. mls)

Based on Grossling's estimates the EUR for Indonesia will be

(A) EUR $= (70 \text{ to } 200) \times 10^3 \times 1.245 \times 10^6$

for oil $= 87$ to 245 billion barrels

and

EUR $= (400 \text{ to } 1,000) \times 10^6 \times 1.245 \times 10^6$ cu. ft

for gas $= 498$ to $1,245$ trillion cu. ft

Based on Akil and Nayoan's estimates of the prospective areas the EUR for Indonesia will be

(B) EUR $= (70 \text{ to } 200) \times 10^3 \times 1.197 \times 10^6$

for oil $= 84$ to 239 billion barrels

and

EUR $= (400 \text{ to } 1,000) \times 10^6 \times 1.197 \times 10^6$

for gas $= 479$ to $1,197$ trillion cu. ft

The undiscovered recoverable oil resources (R_3) of Indonesia can now be arrived at by deducting the sum of the cumulative past production and expanded proved reserves from the estimated ultimate recovery, as given in the equation

$$R_3 = EUR - (CP + R_2)$$

Based on (A) above, and where $R_2 = 1.5R_1$, the undiscovered oil resources (R_3) are

R_3 = (87 to 245) − (8.5 + 14.4) billion barrels

= 64 to 222 billion barrels

Based on (B) above, and where $R_2 = 1.5R_1$, the undiscovered oil resources (R_3) are

R_3 = (84 to 239) − (8.5 + 14.4) billion barrels

= 61 to 216 billion barrels

The oil situation in Indonesia is summarized in Table 7.7 below. Even taking only the lower figures in these ranges Indonesia is estimated to have between 61 to 64 billion barrels of undiscovered recoverable oil.[1]

The estimates of undiscovered oil are based on the amount of oil

TABLE 7.7
THE OIL SITUATION IN INDONESIA, 1980

	Billion Barrels
Cumulative production (1 January 1980)[1]	8.5
Proved reserves (R_1)	9.6
Expanded proved reserves (R_2), where $R_2 = 1.5R_1$	14.4
Estimated ultimate recovery (EUR), based on Grossling's figures for prospective areas	87 to 245
Estimated ultimate recovery (EUR), based on Akil and Nayoan's figures for prospective areas	84 to 239
Undiscovered recoverable oil (R_3), where $R_2 = 1.5R_1$ and where EUR is calculated from Grossling's figures for prospective areas	64 to 222
Undiscovered recoverable oil (R_3), where $R_2 = 1.5R_1$ and where EUR is calculated from Akil and Nayoan's figures for prospective areas	61 to 216

[1]The cumulative production up to 1 January 1980 is obtained by adding the six months (July to December 1979) production in Indonesia (Direktorat Jendral Minyak dan Gas Bumi, *Petroleum and Natural Gas Industry of Indonesia,* July to December 1979 Reports) to the 8.207 billion barrels produced up to 1 July 1979 (*Oil and Gas Journal,* 31 December 1979).

[1]A similar estimate of gas resources has not been attempted because figures on the cumulative production of gas are not available.

expected to occur per unit area of prospective basins, which in turn is based on the average observed outcome from continental size blocs. The size of the prospective area is an important factor in resources evaluation, as the larger the undrilled prospective area the greater the chances of hydrocarbon accumulations discoveries, even when experience has indicated that only a small percentage of such an area actually contains these accumulations (Grossling, 1976). However, since a high percentage of the world's oil has been found in a few 'bonanza' fields[1] whose occurrence is statistically unpredictable, estimates of undiscovered oil based on prospective areas or basin richness must be viewed with great circumspection.

Indonesian geologists of PERTAMINA have recently conducted an assessment of the undiscovered recoverable hydrocarbon resources of Indonesia (Nayoan, *et al.*, 1979; Hariadi, 1980; Suardy, *et al.*, 1980). In their estimations they employed the following parameters:

1. Cumulative production from oil and gas fields
2. Remaining reserves
3. Prospect inventory
4. Recovery factor
5. Success ratio
6. Chance factor, and
7. Possible stratigraphic traps.

In the initial estimations Nayoan *et al.* (1979) delineated forty basins in Indonesia. In a large basin with varying thicknesses of sediments, only those parts with more than 1 000 m of sediments are considered, so that the large basin may be split into two or more basins on this basis. Following Halbouty & Moody (1979) these basins are then classified into (1) intensively explored basins, of which there are 10; (2) partially and moderately explored basins, also 10; and (3) unexplored basins, of which there are 20. The location of these basins is shown in Figure 4.7. No commercial discoveries have yet been

---

[1] In 1968 for example, 70 per cent of the production from 18 major oil-producing regions in the world came from only 20 per cent of the fields. In Indonesia 50 per cent of the production came from only 6 large fields (in 1979). See Van Meurs (1971), Table III, p. 18.

made in the partially explored basins. The unexplored basins are mainly in unfavourable locations — in deep water or in remote areas.

Using the seven designated parameters the total hydrocarbon content per unit volume of sediments in known basins is first determined. The volumetric yield in barrels/acre foot is then applied to estimate the undiscovered recoverable resources of the adjacent prospective structures. The ultimate recoverable supply of hydro-carbon per unit volume derived from the known basins is then used to evaluate the ultimate recoverable hydrocarbon resources in the unknown basins, taking into account chance factors such as the presence of structures, reservoirs and oil generation.

In their preliminary findings presented in 1979 Nayoan and his colleagues took into consideration a number of constraints, including the inclusion of only large gas fields in their calculations. The total remaining ultimate hydrocarbon recovery for Indonesia obtained were as follows:

Oil	30.879	billion barrels of oil (BBO)
Gas	11.095	billion barrels of oil equivalent (BBOE) or approximately 60.825 trillion cu. ft (TCF) of gas
Total	41.974	billion barrels of oil equivalent (BBOE)

A re-assessment of the situation was made by Suardy, *et al.* (1980), taking into account the small as well as large gas fields and the possibility that stratigraphic traps, deeper pooltests and enhanced recovery from low grade deposits could add up to 30 per cent to the ultimate recoverable hydrocarbon supply of known basins. The steps taken in arriving at the revised remaining ultimate hydrocarbon recovery for Indonesia are as follows:[1]

KNOWN BASIN (example)

Average sediment thickness (km)	3.0	(1)
Area (km² × 1,000)	59.418	(2)
Sediment volume (1,000 × km³):		
(1) × (2) =	178.254	(3)

[1] Based on Suardy, *et al.* (1980) and on discussions with Mr Suardy in Jakarta in July 1980.

Cumulative production: oil	1.316	(4)
gas	0.287	(5)
Remaining discovered reserves: oil	0.236	(6)
gas	0.381	(7)

Total ultimate reserves:

$$(4) + (5) + (6) + (7) = \qquad \underline{2.22} \qquad (8)$$

Recoverable undiscovered oil
in known structures, based on
overall recovery factor of 40 per cent $\qquad 4.4 \qquad (9)$
As above, of undiscovered gas,
converted to BBOE on basis of
1 barrel = 6,000 cu. ft $\qquad \underline{0.17} \qquad (10)$

Total $\qquad \underline{4.57} \qquad (11)$

Success ratio derived statistically
from known basins $\qquad 0.135 \qquad (12)$
Recoverable undiscovered resources
from undrilled structures:

$$(11) \times (12) = \qquad \underline{0.617 \text{ BBOE}} \qquad (13)$$

Resources in stratigraphic traps:
$$\frac{30\,((8) + (13))}{100} = \frac{30\,(2.22 + 0.617)}{100} = \qquad \underline{0.851 \text{ BBOE}} \qquad (14)$$

Ultimate recoverable supply of
hydrocarbons in known basin:
$(8) + (13) + (14) = \qquad \underline{3.688 \text{ BBOE}} \qquad (15)$

Ultimate recoverable supply of
hydrocarbons per unit volume:
$$\frac{(15)}{(3)} = \frac{3.688}{178.254} = \qquad \underline{0.021 \text{ BBOE}} \qquad (16)$$

Remaining recoverable reserves
in undiscovered prospects:
$(15) - (8) = 3.688 - 2.22 = \qquad \underline{1.468 \text{ BBOE}} \qquad (17)$

Remaining ultimate recoverable resources:
per cent oil = cumulative production of oil (4)
plus remaining discoverable reserves
of oil (6) divided by total ultimate
reserves (8)
$$= \frac{1.316 + 0.236}{2.22} = \qquad \underline{0.699} \qquad (18)$$

Remaining ultimate recoverable
  oil resources   = (17) × (18)
               = 1.468 × 0.699 =   **1.026 BBO**   (19)

Remaining ultimate recoverable
  gas resources   = $((17) - (19)) \times 6\,MCF^{1}$
               = $(1.468 - 1.026) \times 6\,MCF$
               = $0.442 \times 6\,MCF$ =   2.652 TCF   (20)

UNKNOWN BASIN (example)

Average sediment thickness (km)	2.5	(1)
Area (km² × 1,000)	118.030	(2)
Sediment volume (km³ × 1,000):		
(1) × (2) =	295.075	(3)

Ultimate recoverable supply of hydrocarbons
  per unit volume, derived from known basin and
  applied to unknown basin after geological
  adjustment (see (16) in Known Basin)   0.021 BBOE   (4)

Geological adjustment involving chance
  factors as to:
  presence of structures[2]   0.6   (5)
  reservoirs[2]   0.6   (6)
  environment[2]   0.6   (7)

The adjustment involves multiplying each
  of these factors by the ultimate
  recoverable supply of hydrocarbons per
  unit volume:
  (5) × (6) × (7) × (4)
   = 0.6 × 0.6 × 0.6 × 0.021   0.004536 BBOE   (8)

Estimated ultimate recoverable resources:
  (8) × (3)
   = 0.004536 × 295.075   1.338 BBOE   (9)

As this is an unknown basin, the total ultimate
  recoverable supply and the remaining recoverable
  resources are the same as the estimated
  ultimate recoverable resources, viz.   1.338 BBOE   (10)

---

[1] Conversion factor 1 barrel of oil = 6 MCF of gas.

[2] If from geological assessment the unknown basin has the same possibility of oil generation as an adjacent known basin, each of these factors would be 1. In the example used, these factors are given as 0.6 each, or a 60 per cent chance.

To obtain the ultimate recoverable hydrocarbon
resources, the remaining recoverable resources
are multiplied by the per cent oil (see (18) in
Known Basin):

Oil	1.338×0.699	0.935 BBO	(11)
Gas	$(1.338 - 0.935) \times 6$ MCF	2.418 TCF	(12)

Using the above assessment method and PERTAMINA production
statistics, prospect inventories, drilling and discovery statistics,
reservoir and oil field data both published and unpublished, Suardy, *et
al*., arrived at the following figures for Indonesia:

Remaining Ultimate Recovery

Oil	43.484	billion barrels
Gas	46.198	billion barrels of oil equivalent (BBOE) or approximately 277.503 TCF
Total	89.682	BBOE (see Appendix B)

Their distribution by regional tectonic setting is as follows:

Open Shelf basins	65 per cent (57.9 BBOE)
Foreland basins	10 per cent (9.3 BBOE)
Outer Arc basins	9 per cent (7.7 BBOE)
Interior Cratonic basins	5 per cent (4.8 BBOE)
Unspecified basins	11 per cent (10.0 BBOE)

Of the total remaining ultimate recovery, 22.719 BBOE or 25 per cent
are in water deeper than 200 m. It should be noted that while offshore
production today is being planned—in the North Sea and the Gulf of
Mexico, for example—for water depths of over 300 m, the official
assumption, as used in the Law of the Sea negotiations and adopted by
many governments, is that oil lying in water deeper than 200 m should
be excluded from the category of current producibility. If this
assumption is adopted for the Indonesian situation, then the remaining ultimate recovery of 89.682 BBOE would be reduced to 66.963
BBOE.

Gage and Wing (1980) of the Advance Exploration Group,
CONOCO Incorporated, have recently computed estimates of
recoverable oil reserves for South-East Asia based on a probability/
analogue methodology. The basinal area of this region (extending
N-S from Taiwan to southern Indonesia and E-W from Burma to

New Guinea) is nearly 900,000 sq. mls or 2.33 million sq. km. The basins are classified on the basis of fundamental plate margin interference/interaction processes. Sixty-three basins are identified, which are further grouped into eleven basin-families. Applying statistical data derived from known basins to look-alike but less-known basins, the probabilities of various kinds of exploration outcomes for the lesser-known basins are predicted. From known model basins the authors derived values for 'barrels of oil per square mile of oil-fairway' and 'percentage of total basin area which is inherently oil-fairway'.[1]

The ultimate oil reserves for South-East Asia arrived at by Gage and Wing totalled 69.7 BBO, of which nearly 35.5 BBO have already been found, leaving 34.2 BBO to be discovered. Although the estimates are not specifically on Indonesia but cover a South-East Asia which includes Taiwan and New Guinea, they nevertheless provide an indicator of the remaining undiscovered resources in Indonesia, which must necessarily be less than these estimates. On the assumption that half of these are in Indonesia, then the remaining undiscovered oil resources in Indonesia would be 17 BBO.

The various estimates of the proved reserves and undiscovered resources of oil and gas in Indonesia are given in Table 7.8. Those whose estimates are on the upper end of the scale may cite the following combination of factors for their optimism:

1. The large prospective area of Indonesia.

2. The fact that only 20 per cent of Indonesia has been subjected to detailed geological study, and large areas await intensive exploration.

3. The fact that it is still a poorly researched area, with only 20,000 wells drilled as compared to 2,400,000 wells drilled in the U.S. (PERTAMINA, 1979, p. 111). One-third of the Tertiary basins in Indonesia have not been drilled (Huffington, 1979).

4. Historically, throughout the world, estimates of oil reserves have been proved to be conservative.

5. Apart from sandstone, which was the only reservoir rock in Indonesia up to 1968, two other rock types—carbonates and volcanic clastics—have been proved to be reservoir rocks. All three types of

---

[1] The oil-fairway concept recognizes that oil fields in any given basin tend to be concentrated in trends, and the sizes of such fairways are related to basin type.

TABLE 7.8

ESTIMATES OF THE HYDROCARBON RESERVES AND RESOURCES OF INDONESIA

Source and Date	Estimated Proved Reserves		Remaining Undiscovered Resources		
	Oil (BBO)	Gas (TCF)	Oil (BBO)	Gas (TCF)	Total (BBOE)
Oil and Gas Journal (31 December 1979)	9.6	24			
Ocean Industry (October 1980)	14.0				
World Oil (15 August 1978)	10.985	22.5			
Caltex Pacific Indonesia (1979)[1]	8 to 12				
U.S. Geological Survey (1979)[1]	10.4				
Indonesian Ministry of Mines and Energy (1979)[1]	10 to 15				
Petroleum Economist (August 1980)		37			
Nayoan, et al. (1979)			30.879	11.095	41.974
Suardy, et al. (1980)			43.484	46.198	89.682
Gage and Wing (1980)			17.0[2]		
Estimates based on Grossling's (1976) methodology			61 to 216[3]		
Pertamina (1979)[4]		34	200[5]		
Samadikun (1980)			50[5]		

[1]Cited in U.S. General Accounting Office (1979), p. 1.

[2]Based on the assumption that half of the estimated 34 billion barrels of oil left in South-East Asia will be found in Indonesia.

[3]This is the lower of the two ranges (see Table 7.7 above).

[4]Cited in *Pertamina Today* (1979), p. 111.

[5]Based on a 25 per cent recovery of the oil-in-place.

reservoir rocks occur widely in Indonesia. The search for oil can now be extended over a larger geographic area.

6. The high success rate for oil exploration in Indonesia.

7. The economics of exploration and development which have become increasingly favourable with the steep increases in oil prices, to over $30 per barrel.

However, in so far as the number of wells drilled is concerned, it should be noted that the large number of wells drilled in the U.S. should not be taken as a standard of comparison for Indonesia, as the patterns of ownership of mineral rights in these two countries are different. In the U.S. mineral rights are usually held by the owner of the surface rights. This has provided the incentive for these (mostly private) owners to drill for oil in areas considered marginal elsewhere, and accounts for the large number of wells drilled as well as the large number of small fields (Ivanhoe, 1979; G. L. Shepherd, private communication). In contrast the mineral rights in Indonesia belong to the state, which contracts out the prospective areas to foreign oil companies, which in turn receive a percentage of the oil returns. The drilling pattern must necessarily differ from that in the U.S. in that the returns to investment are different and companies would not be inclined to risk their capital by digging wells thought to be too difficult or to dig wells in areas considered marginal. However much rising oil prices increase the economics of exploration, the drilling intensity in Indonesia will not approach that of the U.S., given the fundamental differences in ownership rights.

A significant number of fields discovered in Indonesia in the last decade had been considered marginal and uneconomic to develop either because the high initial capital cost of the production installation could not be amortized over the expected life of the field, or because not enough was known of the life of the field to justify installing such equipment.

Milton and Sumatri (1978) have classified such marginal fields as:

1. those in remote locations with reserves of 4 to 10 million barrels of recoverable oil and which lie in water up to 100 m deep;

2. those relatively close (less than 17 km) to existing installations, with reserves of 2 to 4 million barrels; and

3. those less than 3 km (2 mls) from an existing facility, with recoverable reserves of only 1 to 2 million barrels of oil.

The steep increases in crude oil prices are now leading to a reassessment of these fields as the economics of development have become more favourable, just as the economics of exploration (which cannot really be separated from development) have also become more favourable and will serve to spur further the search for oil in Indonesia.

In any discussion of the undiscovered petroleum resources of a country it is possible to arrive at widely different answers (as in the case for Indonesia) because of differences in the assumptions, parameters and methodologies employed. Ivanhoe (1979), for example, believes Grossling to be 'super-optimistic' in calculating his world resource figures. The area of prospective basins in the less developed countries as given by Ivanhoe covers only half the area given by Grossling. Furthermore Grossling's areas are multiplied by what Ivanhoe considers to be an exaggerated barrels/sq. ml factor to arrive at estimated ultimate recoveries which, in Ivanhoe's view, are several times too high. This view should be borne in mind when considering the estimates of unrecovered recoverable oil in Indonesia made on the basis of Grossling's methods of estimation.

The estimations of the undiscovered petroleum resources of Indonesia could also be approached, given the availability of data, from angles other than those adopted in this section.[1] But, as one petroleum geologist has said 'About the only thing any estimator can say with certainty about his estimate is that it is wrong. There is simply too much uncertainty in all the approaches to allow the kind of accuracy of estimation that can be achieved for the already discovered petroleum resources' (Sheldon, 1977, p. 988).

For this reason Ivanhoe, a senior geologist of considerable experience, has emphasized the need to treat resource estimates cautiously, and to make a careful distinction between reserves (which he defines as what the engineers know they can produce at a profit, from known fields, with known technologies, in a known time) and resources (which he defines as theoretical estimates of all the oil and

[1]For a discussion of the methodologies of estimating undiscovered petroleum resources, see White & Gehman (1979) and Ion (1980).

gas which may exist in an area, but most of which are unlikely to be converted into reserves) (Ivanhoe, 1979).

Nevertheless, in spite of the hazards and uncertainties involved, resource estimates in Indonesia, as elsewhere, are useful in the assessment of resource adequacy in the formulation of policies and development strategies, and in identifying opportunities for further discoveries, if only on a macro-scale.

VIII

The Role of Oil

BEFORE the Second World War Indonesia's main mineral exports were oil and tin. Together they contributed an annual average of about 29 per cent of the total gross foreign exchange earnings in the period 1936–40, with oil alone contributing 20 per cent (Ter Braake, 1944). After the war foreign exchange earnings from oil increased at a slow rate, reaching 32 per cent of the total gross foreign exchange earnings in 1966. Increased production, in part derived from offshore oil, raised these earnings to 47 per cent in 1973. The quadrupling of oil prices in late 1973 resulted in a dramatic jump in oil revenues, pushing the gross foreign exchange earnings from oil to a record total of $5.1 billion in 1974 as against $1.7 billion in 1973. The contribution of oil to total gross foreign exchange earnings increased to 71 per cent in 1974. Since then oil has contributed an average of 70 per cent to the gross foreign exchange earnings of Indonesia and more than half of its net foreign exchange earnings (Table 8.1).

Oil has also become a major source of domestic revenue. In 1966 government revenue from the petroleum industry amounted to only $2 million or 5 per cent of the total domestic revenue. By the fiscal year 1974–5 this has increased to $2.3 billion or 55 per cent of the total domestic revenue. Thereafter the contribution of tax earnings derived from oil activities has fluctuated between 51 and 56 per cent, but the absolute receipts from oil have increased steadily to a high of $4.8 billion in the fiscal year 1978–9 (Table 8.2). Part of the increases during this period were a result of revisions in the profit-sharing arrangements with foreign oil companies.

The other component of the budgetary income derived from oil is the sale of petroleum products in the domestic market. As Table 8.2

TABLE 8.1

NET FOREIGN EXCHANGE EARNINGS FROM OIL AND NON-OIL EXPORTS, INDONESIA

(US$ Million)

Fiscal Year	Oil(%)	Other Exports (%)	Total
1971−2	204 (21)	784 (79)	988
1975−6	3,138 (63)	1,873 (37)	5,011
1976−7	3,710 (56)	2,863 (44)	6,573
1977−8	4,445[1](56)	3,507 (44)	7,952
1978−9	4,200[1](54)	3,586 (46)	7,786

Source: American Embassy (various years; Jakarta).

[1] Includes net LNG revenue.

shows the receipts from this source have decreased since 1973−4, reflecting governmental policy to subsidize certain petroleum products consumed domestically, especially kerosene which is widely used by the lower income groups. Such subsidies have resulted in a decrease of 1 per cent in the oil-derived budgetary revenue in the fiscal year 1978−9.

In all, the oil sector accounted for about 15 per cent of the per capita GNP in 1978 (American Embassy, 1979).[1] Oil has a vital role to play in maintaining Indonesia's stability and constitutes a major component of the country's economic structure. Oil revenue growth is necessary if the allocation of resources for development is not to be constrained. The Indonesian economy does not have the flexibility and resilience at present to compensate for any significant decline in oil revenue.

The World Bank, in its 1979 Annual Report, projected that Indonesia's net earnings from oil and gas will rise steadily in the first half of the 1980s to total $16.8 billion by 1985. These projections were based on a crude oil price averaging $32 in 1980, an increase in the international oil price index from a 1979 base of 100 to 238 in 1985, a steady growth in LNG earnings to $2.5 billion by 1985 (*Petroleum News*

[1] The per capita GNP was $370 in 1979. It should be noted that with a population of 143 million, every billion dollar of oil earnings will raise the per capita income by only $7.

TABLE 8.2

REVENUE FROM OIL AS A PERCENTAGE OF TOTAL INDONESIAN GOVERNMENT REVENUE, 1966–1979

(US$ Million)

	1966	(%)	1973–4	(%)	1974–5	(%)	1975–6	(%)	1976–7	(%)	1977–8	(%)	1978–9	(%)
Corporation tax on oil companies	n.a.		836	(36)	2,345	(56)	3,009	(56)	3,902	(55)	4,648	(55)	4,982	(52)
Oil product receipts from domestic sales	n.a.		91	(4)	−38	(−1)[1]	Nil		38	(1)	Nil		−142	(−1)[1]
Total petroleum revenue	2	(5)	927	(40)	2,307	(55)	3,009	(56)	3,940	(56)	4,648	(55)	4,840	(51)
Total non-petroleum revenue	34	(95)	1,426	(60)	1,900	(45)	2,393	(44)	3,062	(44)	3,803	(45)	4,584	(49)
Total government revenue	36	(100)	2,353	(100)	4,207	(100)	5,402	(100)	7,002	(100)	8,451	(100)	9,424	(100)

Source: American Embassy, (various years;.Jakarta).

[1] Government subsidy on domestic product petroleum sales.

projected the production of LNG to increase from 8.4 million tons in 1980—2 to 17 million tons in 1985), and an increase of crude oil production from 565 million tons in 1980 to 670 million tons in 1985 (1.83 million barrels/day). These projections of oil production are markedly lower than those in the Third Five-Year Plan of 1979—80 to 1983—4, which sets production to reach 1.83 million barrels a day in 1983—4.

The 18.6 per cent increase in oil production between 1980 and 1985 forecasted by the World Bank is a modest one, and the anticipated increase in oil earnings is expected to arise from increases in oil prices rather than from production. This, in fact, was the case in 1979, when in spite of a 7 per cent fall in oil exports Indonesia recorded a 91 per cent increase in oil revenues over those in 1978 (*Petroleum Economist*, June 1980).

The foreign exchange earning role which Indonesia assigns to its petroleum resources is being eroded through rising domestic consumption of such resources. Improvements in the transportation and distribution infrastructures, and the growing demand for energy from industry and from households have led to a steep growth in the use of oil products. The other important factor is the unbalanced energy-mix usage structure in Indonesia, which is heavily weighted towards oil, natural gas and liquefied petroleum gas (Table 8.3; see also Amin, 1976).

The domestic consumption of oil and oil products, especially kerosene, has more than doubled between 1970 and 1975 (Amin, 1976) and has been increasing at an average of 13.9˙ per cent per annum up to 1979 (PERTAMINA, cited in *New Nations*, 14 February 1980). Part of such increases is due to the under-pricing of oil products through domestic fuel subsidies which have led to what the World Bank has termed 'economic distortions' (cited in *Asia Research Bulletin*, 31 July 1980), such as industries basing their investment decisions on energy costs which are less than its real costs. The subsidy problem has been compounded by a shortage of refining capacity, which has necessitated the import of high-cost refined products.[1].

[1] For technical and price reasons, Indonesia imports Saudi Arabian oil for local refining, while exporting its own low-sulphur high value oil. In 1979 Indonesian refineries supplied 54 per cent of the oil products consumed locally. Enlargement of the

The domestic fuel subsidies have increased from 65 billion rupiahs in 1977−8 to 197 billion rupiahs the following year and to 535 billion rupiahs in 1979−80 (*Pertamina Bulletin*, May 1980). In an effort to contain the problem the Indonesian government raised the prices of all important oil products in April and May 1979 so that of the eight petroleum products, only four—kerosene, diesel, industrial diesel and fuel oil—would continue to be sold at subsidized prices. Nevertheless the net budgetary subsidies, even after the price increases, was calculated to be about 347 billion rupiahs for 1979—80 (*Kompas*, 14 April 1979); in actual fact they came up to 535 billion rupiahs in that fiscal year.

It is evident that such subsidies would be a heavy burden on the economy. In May 1980 a further 50 per cent increase was announced for all the important petroleum products. Kerosene, used by most households for cooking and lighting, went up in price from 25 rupiahs to 37.50 rupiahs per litre (*Pertamina Bulletin*, May 1980). The public reaction, especially to the kerosene price increase was, however, so adverse that the government subsequently reduced the level of price increases to an average of less than 20 per cent (*Straits Times*, 10 July 1980).

The dilemma which Indonesia faces over the rapid growth of energy (mainly oil and gas) consumption[1] and the consequential increase in oil subsidies remains unresolved. The 1981−2 budget, for example, allows for an 84 per cent increase in the government subsidy for oil, from $1.3 billion to $2.4 billion, with a governmental assurance that there would be no increase in the domestic price of oil during the financial year (*Straits Times*, 7 January 1981). As stated by the World Bank, domestic oil pricing remains a key fiscal issue, and unless domestic oil consumption can be lowered to a growth rate of about 10 per cent a year, the net budget subsidies on oil would absorb about 30 per cent of the corporate tax on oil by 1983 (*Asia Research Bulletin*, 31 July 1980).[2]

---

refining capacity when completed in 1986 will enable Indonesia to meet most of its domestic needs for oil products.

[1] Energy consumption per capita increased from 130 to 237 kg of coal equivalent (or 82%) during the period 1960−79 (World Bank, 1981).

[2] In a move to reduce the net budget subsidies on oil, Indonesia raised the domestic

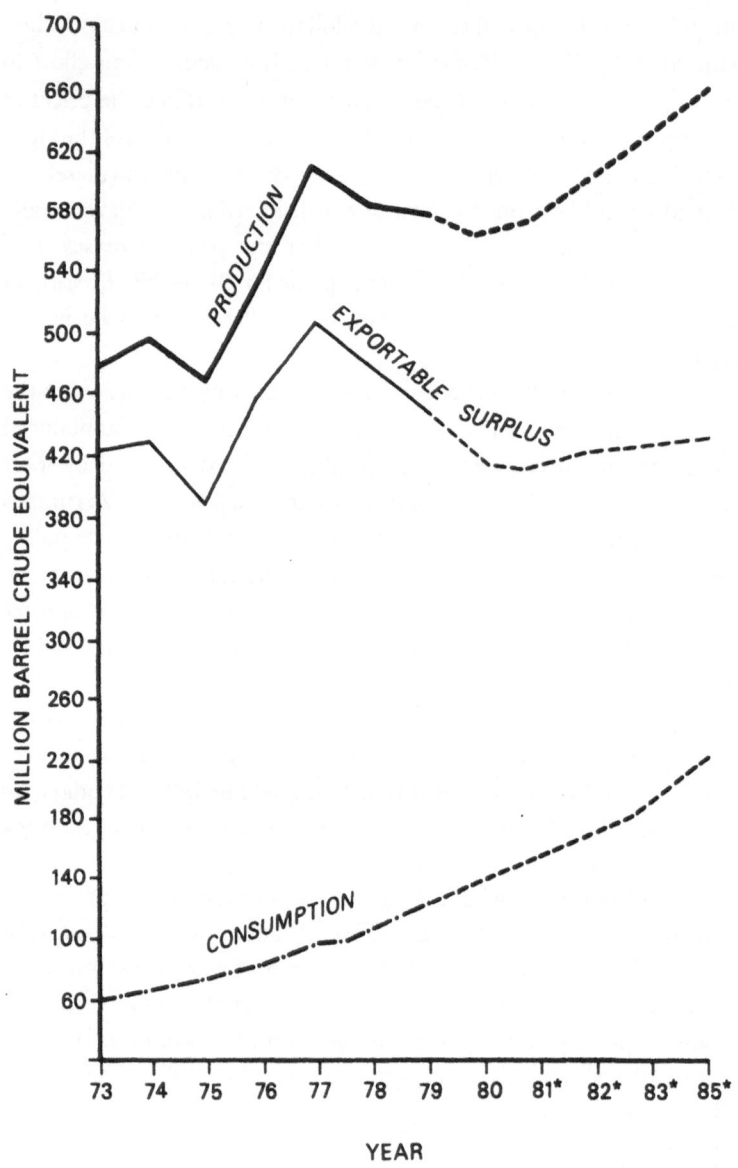

Fig. 8.1   Exportable oil surplus, Indonesia, 1973–1985

Source: World Bank
*World Bank Projection

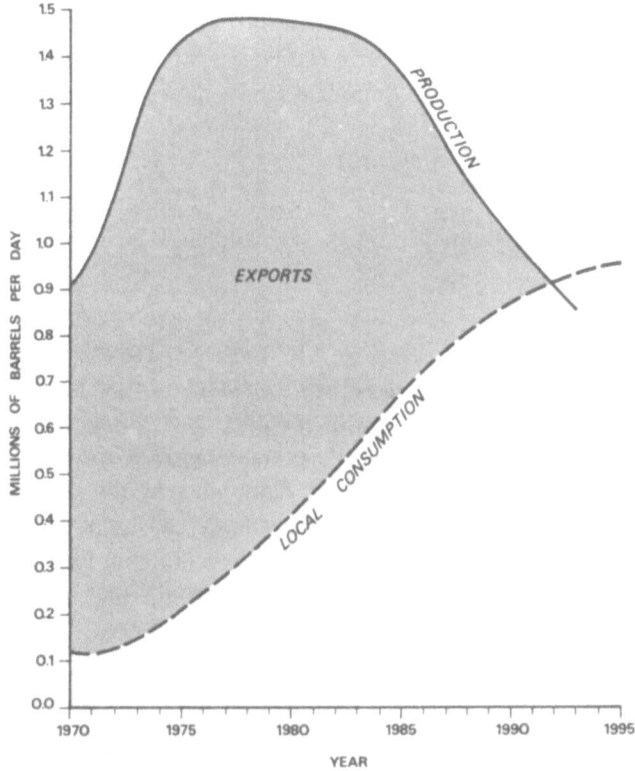

Fig. 8.2   The likely effect of domestic consumption on export
availability of oil in Indonesia
Source: Al-Janabi, April 1977.

In addition, the exportable surplus—the difference between crude
oil exports and crude oil imports and refined petroleum products
imports—would, after a marginal increase between 1981 and 1985,
decline thereafter (Fig. 8.1). Another scenario, drawn up earlier by
the then Senior Economic Analyst of OPEC, showed how internal
demand could reduce Indonesia's exportable surplus to zero by 1992
if no new reserves were added and production continued to fall (Fig.
8.2; see Al-Janabi, 1977; Wijarso, 1977).

prices of petrol, kerosene and other oil products by an average of 60 per cent with effect
from 4 January 1982 (*Straits Times*, 5 January 1982).

Such 'doomsday' scenarios have stimulated the government to review the dual role of oil—as a domestic energy source and as the most important foreign exchange earner—in the economic development of Indonesia, and to formulate a national energy policy. The first energy policy was drawn up in 1977 and used as an input for the Third Five-Year Plan (1979—84). A Ministry of Mines and Energy was established in 1978, which reviewed and updated the energy policy, completing it in March 1980 (Samadikun, 1980; see also Kral, 1978).

The basic objective of the national energy policy is to shift the economy from a mono-energy to a poly-energy dependent one, by strengthening the position of exportable energy sources (mainly oil and LNG) as foreign exchange earners and developing non-exportable energy sources to fuel the economic growth of the country. The targets of the Third Five-Year Plan, whereby the percentage contribution of oil to total domestic energy consumption would gradually be reduced while the contributions of natural gas (apart from LNG), coal, hydroelectric power and geothermal power are increased, are a reflection of this policy (see Table 8.3).

TABLE 8.3
ESTIMATED CONSUMPTION OF COMMERCIAL ENERGY IN INDONESIA, 1978−9 TO 1983−4

Energy Source	1978/79	1979/80	1980/81	1981/82	1982/83	1983/84
	Percentage					
Oil	82.2	79.7	79.6	79.8	79.6	77.7
Natural Gas & LPG	15.8	18.2	18.1	17.8	17.8	17.6
Sub-total	98.0	97.9	97.7	97.6	97.4	95.3
Coal	0.7	0.6	0.9	1.1	1.4	2.5
Hydro-electricity	1.3	1.5	1.4	1.3	1.2	2.2
Geothermal[1]	—	—	—	—	—	—
Total	100.0	100.0	100.0	100.0	100.0	100.0

Source: based on Samadikun, 1980, Table 1.

[1] Power from geothermal sources is projected to increase from 7,000 TCE in 1980−1 to 15,000 TCE in 1983−4.

These sources of energy are classified as 'commercial' energy. In the Indonesian context the traditional non-commercial energy sources such as firewood, charcoal and agricultural waste are still very important. However, no reliable statistics are available on the actual contribution of these non-commercial energy sources, but they were estimated to have made up nearly half of the total energy consumed in Indonesia in 1976 (Samadikun, 1980). These sources of domestic energy will continue to be important, especially in the rural areas where firewood is the common household fuel. In fact, the price subsidy on kerosene is for the special purpose of reducing the rate of deforestation caused by firewood-cutting (Wijarso, 1977). In the overall energy context the use of non-commercial energy is expected to decline in the future due to scarcity or inconveniency considerations (Samadikun, 1980).

Of the commercial non-oil energy sources, coal is by far the most important. Large deposits of coal have been discovered in West, Central and South Sumatra and in East Kalimantan (Fig. 8.3). They vary from lignite to hard coal, with average calorific values of 4 000 to 7 000 kcal. per kg. Proved reserves stood at 100 million tons in South Sumatra and 42 million tons in Ombilin, while indicated reserves were 110 million tons in South Sumatra, 50 million tons in Ombilin and 50 million tons in East Kalimantan (Seminar on Energy Resources in Indonesia, 1980). Coal was mined as early as 1919 at the Bukit Asam field. In 1941 production reached a peak of 800,000 tons, but declined after the war to 180,000 tons in 1978. The only other mine producing coal today is Ombilin. Only 200,000 tons were used for power generation and a variety of end purposes in Indonesia in 1978. As has been pointed out, the use of coal as a substitute for oil in power generation can transfer to each ton of coal an opportunity cost equivalent to five times the price of oil per barrel, as a ton of coal can generate five times as much power as a barrel of oil (Wijarso, 1977). A recent study undertaken by P. T. Shell Mijnbouw indicates that there may be as much as 18 billion tons of coal reserves in Indonesia.[1] Reserves of such magnitude would make the long-term domestic energy situation in Indonesia considerably brighter.

[1]It should be noted that the parameters used by Shell in calculating these figures have not been published.

Fig. 8.3 The coal deposits of Indonesia
Source: Koesoemadinata, 1978.

The problem in the use of coal, as indeed in the use of other non-oil energy fuels, is that the necessary infrastructure does not exist at present, and its development would not only require a long lead time but considerable amounts of scarce capital resources. Technological and management expertise would also have to be acquired.

Indonesia has a hydropower potential estimated at nearly 31 000 megawatts (MW), distributed among the main islands as follows:

Irian Jaya	9 000	MW
Kalimantan	7 000	MW
Sumatra	6 750	MW
Sulawesi	5 600	MW
Java	2 500	MW (Samadikun, 1980)

The present-day installed capacity is only 650 MW or 2.1 per cent of the potential capacity. Although hydropower is attractive as a pollution-free and renewable energy source for the generation of electricity, its development requires substantial capital investment and calls for a long lead time. In addition the potential sites are in geographically remote locations. For these reasons it has been estimated that the installed capacity will not be more than 5 000 MW by the year 2000 (Notodihardjo, 1975; Wijarso, 1977).

The geothermal potential of Indonesia has been estimated to be between 8 000 to 10 000 MW. The geothermal sites are in the islands of the volcanic island arcs of the Indonesian archipelago, mainly in Java and Madura (5 500 MW), Sulawesi (1 400 MW), and Sumatra 1 100 MW). The other sites are in the smaller islands (Fig. 8.4). On the island of Java, where the demand for energy is greatest, the geothermal potential is as follows:

Dieng	1 050	MW
Bauten	360	MW
Kamojag	230	MW
Salak	100	MW
Pelabuhan Ratu	100	MW

Although interest in geothermal power development was shown by the former Dutch colonial authorities as early as 1926, exploration came to a stop in 1928, and it was not until 1964 that the Indonesian government revived that interest. In 1974 PERTAMINA was assigned the task of conducting geothermal surveys and exploration

Fig. 8.4 The potential goethermal areas of Indonesia
Source: Ministry of Mines, 1969.

in the islands of Java and Bali. It has since discovered proved reserves of 58 MW and probable reserves of 3 150 MW (Hadikusumo, *et al.*, 1976; Samadikun, 1980).[1]

The development of geothermal energy suffers from the same disadvantages as hydropower in so far as site locations and immobility are concerned, with the added disability of it being even more expensive, in terms of each unit of power generated, than hydropower. It is estimated that not more than 400 MW will be generated from this source by the turn of the century (Wijarso, 1977).

The use of natural gas as a domestic energy source has increased in recent years. Associated gas as a by-product of crude oil is processed mainly into liquefied petroleum gases (LPG) and natural gas liquids, with the remaining so-called 'tail gas' being used for industries and households in Cirebon and Jakarta. The use of natural gas as a substitute for kerosene was economically attractive when kerosene price was 18 rupiahs per litre; it is even more so with the recent price increase in kerosene. An eightfold increase in natural gas use for domestic energy spaced over twelve years would release about 150,000 barrels of oil per day for export (Wijarso, 1977). Non-associated gas in South Sumatra is also used as feedstock for fertilizers as well as an energy source. However, the non-associated gas at Arun and Badak is processed into LNG and exported. It thus shares with crude oil the position of being a major source of foreign exchange. The Third Five-Year Plan outlines a situation whereby the natural gas contribution to domestic energy consumption will decrease from 18.2 per cent at the beginning of the Plan (1979–80) to 17.6 per cent at the end (Table 8.3).

Apart from these primary sources of energy—coal, hydropower, geothermal power, natural gas—research is also being conducted on solar and wind energy, the latter traditionally and still important in sea transportation based on small sailing boats. Although the International Atomic Energy Agency (IAEA) has indicated that nuclear power will be an economic proposition by 1985, Indonesia will not embark on a nuclear power programme until 1989 when six

---

[1] Some 200 areas in the Indonesian geothermal belt have not yet been investigated (Seminar on Energy Resources in Indonesia, 1980).

nuclear power plants, each with a capacity of 638 MW, will be built over a ten-year period (Wijarso, 1977; *New Nation*, 25 October 1980).

It is apparent from this brief survey of the alternative energy sources that oil will continue to play a dominant role in the domestic energy scene in the 1980s, as for a number of economic and technological reasons the development of the alternatives can only be achieved with a long lead time, given the availability of capital. At the same time the country will have to continue to depend on oil as its main foreign exchange earner.

It would thus appear that Indonesia will not deviate from its policy of lifting its oil out of the ground as rapidly as possible in order to generate development capital as well as satisfy the domestic demand for energy. The option which Malaysia, for example, has chosen through its National Depletion Policy of gradually reducing its oil output from 300,000 to 250,000–260,000 barrels a day in order to conserve its oil reserves, is not one which is really open to Indonesia, at least in the foreseeable future.

Over a longer time horizon Indonesia, in common with other oil producing developing countries, has to face the problem arising from the non-renewable and hence exhaustible nature of its oil and natural gas resources. The problem takes the form of determining the optimal rate of depletion of these resources as well as the inter-related problem of determining the optimal rate of investment. This question has been the subject of recent analyses and debates by a number of resource economists (see for example, the papers in the Symposium on the economics of exhaustible resources, published in the *Review of Economic Studies*, 1974; Al-Janabi, 1979a & 1979b; Motamen, 1979; Jafar Mansur Saad, 1979–80). The issue is a highly complicated one because of several uncertainties: uncertainty about the actual amounts of such resources available at any given time; uncertainty about future technology and the discovery of substitutes; and uncertainty about the population variable.

In the context of many of the oil producing countries of OPEC, which have single-commodity economies based on oil and which lack other natural resources, oil is the main if not the sole source to generate development capital, maintain adequate levels of employment, and maintain public expenditure. There is a school of thought which holds that the objective in these countries should be to ensure that the depletion rate of their (wasting) oil resources is balanced by an investment process that yields the highest rate of economic development, so that when these resources are finally depleted the countries would have achieved the highest attainable level of economic development, if not self-sustaining economic growth (see Al-Janabi, 1979a). In the words of Dr Jafar Mansur Saad of OPEC (1979-80):

The challenge facing every developing oil producing country is the attainment of self-sustaining economic growth. In other words, the challenge is to raise real national income to the level where the volume of savings generated domestically is sufficient to ensure real growth without further dependence on oil revenues. In this context, conservation refers to the long-term process of integrating the entire oil industry of the producing country within the broad framework of a comprehensive national development plan in order to ensure the orderly achievement of the objective of converting or transforming the depletable hydrocarbon resources into productive social and material capital over a specified period of time, which may be designated 'the development horizon' . . .

The length of the time horizon of development and the shape of the depletion profile differ from one Member Country to another in line with the inevitable differences in respect of a multiplicity of social, political and economic factors, notably population and the size of hydrocarbon reserves and other natural resources. A sparsely populated country with a given amount of hydrocarbon reserves may opt for a longer time horizon and lower initial depletion rates than a densely populated country with equal reserves. Such differences have no conceptual relevance. The theoretical essence of conservation is the continuous exchange of oil (and gas) for real capital assets over the time horizon. Consequently, the rate at which this exchange is effected at any point across this time horizon is a crucial determinant of sound economic planning. Indeed, without reasonable control over this variable, or, at least, the ability to predict its behaviour in the short run, conservation remains a mirage.

The Indonesian situation is different from the other OPEC countries which have sole commodity economies, as in addition to its oil and gas, Indonesia has abundant if yet undeveloped alternative energy resources as well as other mineral resources including nickel, copper, bauxite, iron ore and tin. Its timber resources in Kalimantan and Irian Jaya are still extensive, and in fact timber, rubber, palm oil, tin and coffee exports constitute over 20 per cent of its non-oil exports.

Nevertheless, the problem regarding the depletion rates for its oil and gas remains essentially the same as that of the other oil producing developing countries—how to ensure that these wasting assets are exchanged for real capital assets over the time horizon of development. The resolution of this problem will be the major task confronting Indonesia's planners in the years ahead.

Appendixes

APPENDIX A
CONTRACT AREAS, INDONESIA, JANUARY 1980

Operator & partners	Contract area: present, original size & relinquishments (km²)	Contract signing	Contract particulars	History
Amoco 50.0% Pertamina 50.0	NE Sumatra—Panai (On) Present: 10,495 50% relinquishment in 1987.	9/79	Joint-venture; expenditure $5 million for 3 yr, then Pertamina pays 50%. Signature bonus $1.5 million. Production bonuses $1.5 million at start commercial prod, $1.9 million at start of profit, $2 million at 50,000 b/d, $3 million at 100,000 b/d.	
Amoseas Calasiatic 25.0% Texaco Overseas 25.0 Conoco 25.0 Total 25.0	Irian Jaya in Mimika-Eilanden district Present: 40,293 Original: 53,456	28/10/71	Production-sharing	Previously operated by Conoco.

APPENDIX A (*continued*)

Company / Partners	%	Block / Area	Amount	Date	Type	Remarks
Amoseas		East Kalimantan (On)		6/12/69	Production-sharing	Original contract signed by Shell. Esso farmed in 1974, Chevron and Texaco acquired interest in 1978.
Chevron Kutei	20.0%	Present:	19,193			
Texaco Kutei	20.0	Original:	32,060			
Kaltim Shell	30.0					
Esso	30.0					
Amoseas		Natuna Sea, Block C		12/79	Signature bonus $1.5 million, expenditure $18.5 million 6 yr.	
Chevron	50.0%	Present:	29,396			
Texaco	50.0					
Aquitaine Indonesie	51.0%	N Sumatra (Off)		14/4/78 (2 yr)	Production-sharing	Plascom farmed in 1979.
North Sumatra Oil	44.0	Present:	8112			
Plascom Ltd	5.0	Original:	16,220			
Asamera	60.0%	N Sumatra (On) Bika		1/9/61	Production-sharing	Original contract of work modified 24/6/63, 14/6/67, 25/9/70, 12/10/70.
Union Texas	30.0	Present:	3740			
Benedum Trees	10.0	Original:	2978			
		No relinquishments to date				
Asamera	100%	S Sumatra—Tempino Field (On)		15/10/68 (30 yr)	Technical assistance	Original contract signed by Redco; later acquired as a subsidiary of Asamera.
		Present:	285			
		Original:	285			
		No relinquishment obligations				

Operator & partners		Contract area: present, original size & relinquishments (km²)	Contract signing	Contract particulars	History
Asamera	100%	Corridor Blk, Bentayan Present: 11,260 Original: 11,260	15/10/68	Technical assistance	15/10/68 Redco, farmed-out to Stanvac in Jan 71, returned to Asamera 15/1/77.
Associated Australian Resources (AAR)	100%	Irian Jaya Ceram—Bulu (On) Present: 32,302 Original: 113,500	1/11/69 (30 yr)	Production-sharing; info: $3 million; expenditure: $14.5 million over 8 yr.	1/11/69: Original contract with Gulf & Western; 11/72: AAR acquired 100% working interest.
Atlantic Richfield Phillips Amoseas	25.0% 50.0 25.0	NE Kalimantan (On) Present: 6720 Original: 16,835 Three relinquishments	8/8/71 (30 yr)	Signature bonus: $1 million; info: 4 million; expenditure: $19.7 million over 8 yr, production bonus: $2 million at 50,000 b/d, $4 million at 100,000 b/d.	Amoseas acquired 25% in mid-1976.
Atlantic Richfield Natomas Reading & Bates Ramah Properties	46.0% 36.8 12.2 4.9	NW Java Sea (Off) Present: 27,677 Original: 55,436 Two relinquishments	19/1/67 (30 yr)	Production-sharing	Arco acquired holdings of Sinclair Oil; Reading & Bates bought Carver Dodge in 1969.

APPENDIX A (*continued*)

British Petroleum Gulf	50.0% 50.0	Onshore Sulawesi Present: Orginal:	11,214 29,260	24/10/77 (30 yr)	Production-sharing	Area awarded to Gulf in 1970. BP acquired 50% interest 5/9/74. BP assumed operatorship 1/7/76.
Calex Standard Oil Calif Calasiatic	 50.0% 50.0	Sumatra (On) Present:	48,520	9/63 extensions 8/8/71, 20/1/75	Contract of work; parts production-sharing	Original "contract of work" for central Sumatra area dates to 1931; 1968: six blocks added (42,838 sq km) and 1st relinquishment, leaving 50,764 sq km; 9/8/71: production-sharing contract extensions will apply to additional 22,2201 sq km, effective 28/11/83.
Cities Service Citco Indonesia Petroleum Corp Marathon	50.0% 50.0 50.0	E Java Sea—Siri (Off) Present:	11,180	12/2/79 (10 yr)	Production-sharing: expenditure $1.3 million for 2 yr. Signature bonus $200,000. Production bonuses, $250,000 @ 5000 b/d, $2,273,000 @ 50,000 b/d, $4,545,000 @ 75,000 b/d, $6,818,000 @ 100,000 b/d.	Marathon farmed in 10/79.

Operator & partners		Contract area: present, original size & relinquishments (km²)	Contract signing	Contract particulars	History
Conoco Ultramar	50.0% 50.0	E Kalimantan (Off) Present: 3865	16/1/75 (30 yr)	Production-sharing: signature bonus: $1 million; production bonus: $2 million @ 50,000 b/d. $5 million @ 100,000 b/d, $10 million @ 500,000 b/d.	Area relinquished by Total Indonesie.
Conoco Getty Gulf Inpex	40.0% 25.0 17.5 17.5	S China Sea—Blk B (Off) Present: 31,116 Original: 106,708	16/10/68 (30 yr)	Production-sharing signature bonus: $7 million; expenditure: $14 million over 1st 6 yr; production bonus: $3 million at 50,000 b/d, $3 million at 100,000.	Amoco gained one-third interest in Conoco's Indonesia holdings in late 1972.
Conoco Pertamina Mitsui Oil Inpex	25.0% 50.0 12.5 12.5	Onshore Irian Jaya, Kepala Burung Selatan Block A. Birds Head Present: 9200 to be successively relinquished to 3680 sq km in 8 yr.	10/22/77	First ever 50/50 joint-venture by Pertamina; exploration costs, production to be shared equally, after which contractor's share will be split 85/15. Conoco agreed to spend	6/78 Mitsui and Indonesia Petroleum farmed-in for 12.5% interest each of Conoco's original 50% share.

	Interest	Block	Date	Terms	Remarks
				$15 million on exploration in first three years. Signature bonus $3.15 million, production bonuses $1 million over 50,000 b/d, another $1 million over 150,000 b/d.	
Conoco California Standard Texaco CFP	50.0% 50.0	Irian Jaya-Nanka block Present: 42,734	12/79	Production-sharing; expenditure $8 million in 4 yr.	No drilling in 1979.
Deminex Pertamina	50.0% 50.0	NE Kalimantan-Simenggaris (On) Present: 4285 Original: 4285 Relinquishments in 7 yr to 1170 sq km	9/6/78	Joint prod-sharing; expenditure: $9.5 million in 3 yr; signature bonus $2 million; production bonuses; $1 million @ 25,000 b/d. $2 million @ 75,000 b/d.	
Esso Gulf Sunmark Louisiana L&E	50.0% 30.0 12.5 7.5	W Irian Jaya Bomberai Pen (On) Present: 9660 Original: 14,997	8/8/71 (30 yr)	Production-sharing; signature bonus: $3 million; expenditure: $10.1 million over 8 yr.	24/10/70: Original contract Whitestone Petroleum; 3/71: Gulf acquired 80%; 13/7/72; LL&E acquired Whitestone's 20%; 1976: Sunmark farmed-in for 25%, Esso became operator 28/11/77.

Operator & partners		Contract area: present, original size & relinquishments (km²)	Contract signing	Contract particulars	History
Gulf Oil		Natuna Sea—Block A	22/9/79	Production-sharing; expenditure $80.7 million in 6 yr. Signature bonus $20.1 million. Production bonuses, $10 million @ 50,000 b/d, $50 million @ 100,000 b/d, $100 million @ 200,000 b/d.	
Sumatra Gulf	100.0%	Present: Relinquishments in 6 yr to 5000 km²			
Hudbay Oil	50.0%	Malacca Strait (Off)	5/8/70 (30 yr)	Production-sharing; signature bonus: $4 million (plus info).	5/8/70: Original Kondur contracts; 3/6/71: Arco acquired 51% and operatorship from Pan Ocean; 1/3/79 Hudbay Oil became operator.
Atlantic Richfield	25.50	Present: 15,908			
Bridger Petroleum	1.95	Original: 39,845			
Pan Ocean	17.55	Three relinquishments			
Kondur Oil	5.0				
Huffco (see below)		S Sumatra (On)	8/8/68	Production-sharing; signature bonus: $1 million (includes block in Kalimantan under contract): expenditure: $11 million (for both areas).	Original contract covered all Sumatra south of equator not assigned by Pertamina. Minimum depth clause: 1500 meters. Production rights excluded for proven production areas.
		Present: 5160			
		Original: 5160			
		Relinquishment: At option of Huffco			

APPENDIX A (*continued*)

Huffco					
Roy M. Huffington	10.0%	E Kalimantan (On)	8/8/68 (30 yr)	Production-sharing; signature bonus: $1 million; expenditure: $11 million 1st 6yr, (8-yr explor period)	Minimum depth clause 2230 m; see Huffco—S Sumatra.
Huffington	35.0	Present: 12,617			
Ultramar	35.0	Original: 12,617			
Union Texas	10.0	Relinquishment at option of Huffco			
Australian Oil Universal					
Tankship	5.0				
Virginia International	5.0				
Jambi Oil		S Sumatra—			
Teikoku	18.0%	Jambi A-block (On)	9/6/78	Joint production-sharing; expenditure $16 million in 3 yr; signature bonus: $4 million; prod bonuses $1 million @ 50,000 b/d, $1.5 million @ 100,000 b/d, $2 million @ 150,000.	
Mitsui	6.0	Present: 5535			
Sumitomo	6.0	Original: 5535			
JNOC	70.0	To be reduced to 2500 km^2 by 8th yr			
Jambi Shell	50.0%	S Sumatra: Jambi (On)	12/2/79	Joint-venture	
Pertamina	50.0	Present: 19,030			
Japex Sumatra	50.0%	S Sumatra—Lampung (On)	14/7/78	Joint production-sharing	21/7/78 Transferred from Japex to Japex Sumatra.
Pertamina	50.0	Present: 21,000			
		Original: 21,000			

APPENDIX A (continued)

Operator & partners	Contract area: present, original size & relinquishments (km²)	Contract signing	Contract particulars	History
Marathon Oil 33.3% Hudbay 33.3 Coastal States 33.3 Gas	S China Sea— Kakap Block A (Off) Present: 5165 Original: 5165	3/75 (30 yr)	Signature bonus: $3 million; info: $3 million; expenditure: $20.5 million over 8 yr.	Original area relinquished by Agip. Phillips, Teneco later held one-third shares but gave them up. Coastal States signed 2/78 to spend $2.5 million to drill one well.
Marathon 25% Indonesia Sun Oil 75	Irian Jaya (On) Present: 13,318 Original: 17,760	6/10/73 (30 yr)	Production-sharing; signature bonus: $1.5 million; expenditure: $11.3 million over 1st 8 yr.	Pertamina has option to take 10% in case of commercial discovery. 1978: Marathon took over operatorship from Sun.
Marathon 40.0% Amerada Hess 40.0 Hudbay Oil 20.0	Natuna Sea—B Block (Off) Present: 24,605 Original: 24,605 Area to be successively reduced to 5094 km² at end of 6th year	11/79	Production-sharing. Expenditure $62 million over 1st 10 yrs; $14 million in 1st 2 yr. Signature bonus $8.5 million	

APPENDIX A (continued)

Company	%	Area / Block	Area	Original/Present	Date	Agreement	Notes
Mobil Oil Indonesia	100%	North Sumatra—B Block (On)	3745	Original: 1872 Present:	22/7/68*	Production-sharing	22/7/68*: area acquired by Mobil Oil Indonesia Inc. Production from Arun field commenced 1978.
Mobil Petroleum Indonesia Inc.	100%	Makassar Straits (Off)		Original: 20,914 Present: 13,594	14/3/73 (30 yr)	Production-sharing	
Mobil Exploration Indonesia Inc.	100%	North Sumatra (Off & On)		Original: 20,767 Present: 25,037	16/10/68 (30 yr)	Production-sharing	25/1/70: 3670 km² substitution as compensation for boundary changes.
Mobil Oil Indonesia	100%	Natuna Sea—Block DI & DII		Present: 10,307 Original: 10,307		Production-sharing. Signature bonus: $22 million. Expenditure $140.4 million over 1st 10 yr.	
Mobil Mobil Peusangan Pertamina	50.0% 50.0	N Sumatra—Peusangan (Off) Peusangan	3440		6/79	Joint venture; expenditure $15 million for 3 yr, then Pertamina pays 50%. Signature bonus $1 million, discovery bonus $5 million. $7.6 million on start of profit.	Pertamina's first offshore joint venture.

Operator & partners	Contract area: present, original size & relinquishments (km²)	Contract signing	Contract particulars	History
Natomas (Iiapco) 52.68%	Java Sea (S Sumatra) E & W Blocks (Off) (Gita and Selatan)	6/9/68 (30 yr)	Production-sharing	June 14, 1971: Shell/Iiapco split block at 160° 30'E N'ward from N'western corner of Iiapco-Arco block. Shell commitment of $156 million for 43.6% of Iiapco 56%. ($41 million for development of Iiapco proven production). 1st relinquishment: postponed for $1.25 million fee.
Reading & Bates 10.17	Present: 65,972			
Hudbay 8.66	Original: 132,090.59			
R & B Offshore 5.53	Final relinquishment 9/80 to			
Drlg	15,000 km²			
Sunda Shell 5.00				
Getty Oil 4.97				
Japan Oil				
Sulfur 3.96				
Tidewater 3.76				
Warrior 0.80				
Trans Ocean				
Gulf 0.40				
Tom Brown 0.30				
Petromer-Trend 27.0%	Irian Java—Vogelkop (On)	15/10/70 (30 yr)	Production-sharing; signature bonus: $150,000; expenditure: $5.3 million over 10 yr.	
Eurafrap 15.0	Present: 1288			
C Itoh 10.0	Original: 5158			
S Pacific Oil 10.0				
Moeco 15.0				
Mapco 10.0				
Southern Cross 10.0				
Banque de Paris 3.0				

APPENDIX A (continued)

Company	%	Area / Block	Date	Contract Terms	Notes
Pexamin Pacific Inc	100%	E Java Sea (Off) Present: 15,293 Original: 20,390 25% relinquishment 30/8/77 63% relinquishment due 9/85	3/75 (30 yr)	Production-sharing; expenditure: $11.8 million over 8 yr; production bonus: $1.5 million.	Area relinquished by Cities.
Phillips Agip Conoco	50.0% 25.0 25.0	Irian Jaya—Teluk Berau (On) Present: 15,000 Original: 100,000	6/10/68 (30 yr)	Production-sharing: info: $1.5 million; expenditure: $16 million over 1st 8 yr; production bonus: $5 million at 75,000 b/d, $1.5 million at 100,000 b/d, $2 million at 200,000 b/d.	10/10/68: Original contract with Agip; 4/69: Phillips acquired 50% and operatorship.
Phillips		Irian Jaya—Kamura (On) Present: 22,385	6/79	Production-sharing	
Shell		Irian Jaya—Mamberamo (On) Present: 14,675	12/79	Production-sharing. Signature bonus $25 million. Expenditure $137.5 million over 1st 10 yr.	
Stanvac Exxon Mobil	50.0% 50.0	Japura Block & Kampar Block in C Sumatra; Pendopo and Raja blocks in S Sumatra Present (total): 11,701.8 Original (total): 19,819.4	28/11/63	Contract of work (20 yr)	Kampar block: 1st relinquishment in Nov 1976: 4070.6; 2nd relinquishment in Dec. 1973: 4047 km².

APPENDIX A (*continued*)

Operator & partners	Contract area: present, original size & relinquishments (km²)	Contract signing	Contract particulars	History
Stanvac Exxon 50.0% Mobil 50.0	S Sumatra—Lakitan, Keruh, Lematang & Kebur blocks Present (total): 8456.86 Original (total): 9021	1/70	Contract of work (23 yr)	Lematang block: 1st relinquishment in 1975: 268.148 km²; Kebur block: 1st relinquishment in 1975: 296.091 km².
Stanvac Exxon 50.0% Mobil 50.0	Rimau Block in S Sumatra Present: 2854 Original: 2854	4/73	Production-sharing	
Stanvac Exxon 50.0% Mobil 50.0	C Sumatra—Barisan block Present: 247 Original: 247	4/73	Production-sharing	
Stanvac Exxon 50.0% Mobil 50.0	C Sumatra— Pamai Taluk Block A Present: 1900 Original: 1900	13/10/78	Production-sharing	
Sumatrapex White Sheild 1.3% 98.7	S W Sumatra— Lampung (On & Off) Present: 20,951 25% relinquishment in 8/77	9/8/69 (30 yr)	Production-sharing; signature bonus $75,000 and $0.15 million to Indonesian school system	Original contract with Asia Oil; 1/9/71: White Shield gained 85%; 7/76; Gulf farmed-in for 47.3% and 2 wells but left after 1 well.

APPENDIX A (*continued*)

		Area	Date	Terms	Notes
Tesoro Petroleum Pertamina		NE Kalimantan—Tarakan/Sanga Sanga (On) Present: 582 Relinquishment: None required	15/10/68 1969: amended	Production-sharing; expenditure: $7.5 million over 2nd—5th yr.	15/10/68: Redco; 1969: Contract amended; 4/71: sold to Tesoro.
Total Indonesie Pertamina Muturi Inpex Irian Jaya Pet Co	25.0% 50.0 12.5 12.5	Onshore Irian Jaya, Kepala Burung Selatan Block B (Muturi area) Present: 8890 to be successively relinquished to 4000 km² by end of 8th yr.	1977	Pertamina's 2nd 50/50 joint venture; exploration costs, production to be shared equally, after which Total's share will be split 85/15. Total will spend $23 million on exploration in first 3yr; Prod bonuses: $2 million @ 100,000 b/d, $2 million @ 200,000 b/d.	7/78: Inpex and Irian Jaya farmed-in to Total's share.
Total Indonesie Calasiatic Texaco Overseas Japan Pet Expl	25.0% 25.0 25.0 25.0	Bunyu area Present: 7162 Original: 32,490	31/3/67	Production-sharing	Original award to Japex. Total farmed-in 1970.
Total Indonesie Inpex	50.0% 50.0	E Kalimantan—Mahakam Present: 8685 Original: 16,870	31/3/67	Production-sharing	Original contract with Japex. Total farmed-in 1970.

APPENDIX A (continued)

Operator & partners		Contract area: present, original size & relinquishments (km²)	Contract signing	Contract particulars	History
Union	100%	E Kalimantan—Makassar Strait (On & Off) Present: 5123 Original: 24,485	25/10/68 (30 yr)	**Production-sharing;** signature bonus: $42 million (incl info); expenditure: $4 million over 8 yr, production bonus: $1.5 million @ 75,000 b/d; $1.5 million @ 175,000 b/d.	25/10/68; Contract for 4014 km² (on and off); 21/4/70: extension on-shore of 5666 km² for signature bonus of $0.55 million and expenditure from $4 million to $6 over 1st 6 yr; 6/7/70: Japex relinquished off-shore area adjacent to Balikpapan and Union took 2663 km²; 25/8/70: 1468 km² added.

APPENDIX A (*continued*)

Union Inpex	50.0% 50.0	E Kalimantan—Attaka (Off) Present: 285 (Union) Original: 285 (Japex)	6/7/70 (Union) 31/3/67 (Japex)	Production-sharing	Attaka straddles Japex-Total and Union's Mahakam area blocks; 6/3/67: Japex awarded Mahakam; 31/3/70: Japex 1st relinquishment; 17/4/70: Japex and Union agree on joint development of Attaka; 6/7/70: Union buys two areas relinquished by Japex.
Union Katy Industries	80.0% 20.0	SE Kalimantan—Teweh (On) Present: 17,835 Original: 23,775 Relinquishments: 14/1/80—5918, 14/1/82—3897, 14/1/84—4156.	14/1/74	Production-sharing; signature bonus & info bonus — $10 million.	Original contract Katy. Union Oil Co of Indonesia assigned its rights & interests to Union Oil Co Kalimantan.

Source: Petroleum News, January 1980.

APPENDIX B

SUMMARY OF RECOVERABLE HYDROCARBON RESOURCES IN INDONESIA

Basin Number	Name of Basin and Province	Average Sediment Thickness (km)	Area (km²) x 1000	Total Ultimate Rec. Supply in BBOE	Remaining Ultimate Rec. Supply in BBOE
1	North Sumatra Basin	2.5	78.472	6.964	3.501
	North West Sumatra Province	2.5	31.428	—	—
2	Sibolga Basin	2.5	61.320	1.825	1.825
3	Central Sumatra Basin	1.5	39.345	8.385	2.741
4	Bengkulu Basin	2.5	118.030	1.338	1.338
5	South Sumatra Basin	3	59.418	3.688	1.468
6	Sunda Basin	1.5	23.440	0.525	0.143
7	North West Java Basin	1.75	49.504	2.095	1.126
8	Biliton Basin	1.25	16.763	0.122	0.122
9	South Java Basin	2.5	77.636	1.014	1.0145
10	North East Java Basin	2	26.800	0.559	0.379
11	Pati Basin	1.75	28.036	0.261	0.261
12	North East Java Sea Basin	1.25	59.781	0.675	0.668
13	East Natuna Basin	3	71.927	1.175	0.455
14	West Natuna Basin	2.5	69.454	0.194	0.111
15	Ketungau/Melawi Basin	1	41.672	0.051	0.051
16	Barito Basin	1.5	49.200	0.471	0.3199
17	Asem-Asem Basin	2.5	35.054	3.145	3.145
18	Kutei Basin	4.0	45.939	25.761	22.931

19	Tarakan Basin	3.0	38.145	2.472	2.065
20	Sulawesi Basin	4.0	66.363	0.550	0.550
21	Makassar Strait Basin	4.0	39.913	7.171	7.171
22	Lariang Basin	4.0	29.127	5.227	5.227
23	Makassar Basin	3.0	89.636	12.064	12.064
24	Gorontalo Basin	1.5	110.479	3.485	3.485
25	Banggai Basin	1.5	95.959	3.027	3.027
26	Bone Basin	2.0	58.080	2.442	2.442
27	Sulawesi Tenggara Basin	1.5	70.075	4.244	4.244
28	Flores Basin	3.0	114.570	0.447	0.447
29	Bali Basin	3.0	31.565	0.123	0.123
30	Sawu Basin	3.0	92.802	0.362	0.362
31	Timor Basin	3.0	51.136	0.704	0.704
32	Banda Basin	3.0	377.525	0.526	0.526
33	Halmahera Basin	2.0	133.383	0.290	0.290
34	Waigeo Basin	2.0	123.737	0.429	0.429
35	Salawati Basin	2.0	102.276	1.056	0.655
36	Bintuni Basin	2.5	104.166	0.334	0.334
37	Aru Basin	3.0	162.247	0.844	0.844
38	Waropen Basin	3.0	279.696	1.455	1.455
39	Akimengah Basin	2.5	171.717	0.744	0.744
40	Sahul Basin	2	258.206	0.895	0.895
	TOTAL			107.139	89.682

Source: Hariadi (1980).

Bibliography

ABOLFATHI, F., *et al.* (1977), *The OPEC Market to 1985* (Toronto).

ADAMS, T. D. (1980), 'Petroleum exploration in Southeast Asia: a two-year overview (1978–1979)', *Offshore Southeast Asia Conference* (Singapore) (unpublished).

ADINEGORO, A. R. UDIN (1973), 'Stratigraphic studies by the Indonesian Petroleum Institute (Lemigas)', *C.C.O.P. Technical Bulletin, No. 7* (Bangkok), pp. 55–74.

AKIL, ISMET & NAYOAN, G. A. S. (1973), 'Notes on offshore petroleum development in Indonesia', *C.C.O.P. Report of the 9th Session* (Bangkok), pp. 118–23.

AKIL, ISMET & PEKAR, L. (1977), 'Ten years of petroleum exploration offshore Indonesia', *C.C.O.P. Technical Bulletin, Vol. 11* (Bangkok), pp. 69–96.

AKUANBATIN, H. & ARDIPUTRA, D. (1976), 'Geology of the East Benakat oil field, S. Sumatra', *Proceedings of the 5th Annual Convention, Indonesian Petroleum Association*, pp. 59–68.

ALFORD, M. E., *et al.*, (1975), 'Development of the Arun gas field', *Proceedings of the 4th Annual Convention, Indonesian Petroleum Association*, pp. 173–88.

American Association of Petroleum Geologists (1971), *Origin of Petroleum* (Colorado).

American Embassy (1976), *Indonesia's Petroleum Sector, 1976* (Jakarta).

_____ (1977), *Indonesia's Petroleum Sector, 1977* (Jakarta).

_____ (1978), *Indonesia's Petroleum Sector, 1978* (Jakarta).

_____ (1979), *Indonesia's Petroleum Sector, 1979* (Jakarta).

____ (1980), *Indonesia's Petroleum Sector, 1980* (Jakarta).

____ (1980), *Indonesia. Industrial Outlook: Minerals* (Jakarta), August, mimeo.

AMIN, ADIMIN (1976), 'The role of oil in Indonesia's economic development', *Prisma*, May, pp. 106–13.

ARCE, R. (1961), *Mineral Resources of Indonesia* (U.N. Dept. of Economic & Social Affairs, New York).

ARIEF, SRITUA (1977), *Financial Analysis of the Indonesian Petroleum Industry* (Jakarta).

____ (1976), *The Indonesian Petroleum Industry* (Jakarta).

BAIN, H. F. (1933), *Ores and Industry in the Far East* (New York).

BARNEA, J. (1977), 'The future supply of nature-made petroleum and gas', in Meyer (ed.), pp. 23–40.

BARTLETT, A. G., et al. (1972), *Pertamina: Indonesian National Oil* (Singapore).

BEDDOES, L. R., JR. (1980), 'Hydrocarbon plays in Tertiary basins of Southeast Asia', *Offshore Southeast Asia Conference, SEAPEX Session* (Singapore) (unpublished).

BEERS, H. W. (ed.) (1970), *Indonesia: Resources and their Technological Development* (University Press of Kentucky, Lexington).

BELTZ, E. W. (1944), 'Principal sedimentary basins in the East Indies', *Bulletin of American Association of Petroleum Geologists*, pp. 1440–54.

BEMMELEN, R. W. VAN (1949), *The Geology of Indonesia, Vol. II, Economic Geology* (The Hague).

BRAAKE, A. L. TER (1944), *Mining in the Netherlands East Indies* (New York).

BRAMONO, B. (1974), 'LNG projects in Indonesia', *Proceedings of the 3rd Annual Convention, Indonesian Petroleum Association*, pp. 265–79.

BRODJOUEGORO, SOEMANTRI (1969), 'The policy on oil in Indonesia', *OPEC International Oil and the Energy Policies of the Producing and Consuming Countries* (Vienna), pp. 21–4.

CARLSON, S. (1976), *Indonesia's Oil* (Georgetown University, U.S.A.).

CLEGG, M. W. (1967), 'Secondary recovery methods—today and tomorrow', *World Petroleum*, Vol. 38, pp. 82–6.

Committee for Co-ordinational of Joint Prospecting for Mineral

Resources in Asian Offshore Areas (C.C.O.P.) (1976), *The Offshore Hydrocarbon Potential of East Asia, A Decade of Investigations, 1966–1975* (Bangkok).

COWPER, R. (1980), 'Arun and Bontang: deals in the offing', *Petroleum News*, April, pp. 16–18.

DASH, B. P., *et al.* (1972), 'Seismic investigations on the northern part of the Sunda Shelf south and east of Great Natuna Island', *C.C.O.P. Technical Bulletin, No. 6* (Bangkok), pp. 179–96.

Departemen Pertambangan R.I. (1977), *Pertambangan Indonesia, 1976* (Jakarta).

Directorate-General of Oil & Gas, Indonesia (1975), 'Petroleum exploration and development in offshore areas of Indonesia from January 1974 to May 1975', *C.C.O.P., Proceedings of the 12th Session* (Tokyo), pp. 246–53.

DJALAL, HASJIM (1979a), 'Indonesia and the new extents of coastal state sovereignty and jurisdiction at sea', *The Indonesian Quarterly*, January, pp. 80–93.

———(1979b), 'Conflicting territorial and jurisdictional claims in South China Sea', *The Indonesian Quarterly*, July, pp. 36–52.

ECAFE (1978), *Working Group Meeting on Energy Planning and Programming* (Bangkok).

EMERY, K. O., *et al.* (1972), 'Geological structure and some water characteristics of the Java Sea and adjacent continental shelf', *C.C.O.P. Technical Bulletin, No. 6* (Bangkok), pp. 197–223.

ENGLEHARDT, J. M., *et al.* (1980), 'Gravel pack design, performance and evaluation, Duri field', *9th Annual Convention, Indonesian Petroleum Association* (unpublished).

FALLAH, R. & FESHARAKI, F. (1977), 'The new world economic order: the energy dimension', *OPEC Review*, October, pp. 7–20.

FLETCHER, G. L. & SOEPARJADI, R. A. (1976), 'Indonesia's tertiary basins—the land of plenty', *Offshore Southeast Asia Conference, SEAPEX Program* (Singapore) (unpublished).

Fortune (1973), July.

FROIDEVAUX, C. M. (1977), 'Tertiary tectonic history of the Salawati area, Irian Jaya, Indonesia', *Proceedings of the 6th Annual Convention, Indonesian Petroleum Association*, pp. 199–220.

GAFFNEY, P. D., *et al.* (1976), 'Economic appraisal of the potential petroleum resources of the Asian Pacific region', *Offshore Southeast Asia Conference* (Singapore) (unpublished).

—— & Moyes, C. P. (1978), 'Competitive legislation—the key to Asia Pacific petroleum prospects', *Offshore Southeast Asia Conference* (Singapore) (unpublished).

Gage, M. S. & Wing, R. S. (1980), 'Southeast Asian basin types versus oil opportunities', *9th Annual Convention, Indonesian Petroleum Association* (unpublished).

Gerretson, F. C. (1953), *History of the Royal Dutch*, Vol. 1 (Leiden).

Glassburner, B. (ed.) (1971), *The Economy of Indonesia* (Cornell University Press, Ithaca).

Goldstone, A. (1977), 'What was the Pertamina crisis?', *Southeast Asian Affairs*, ISEAS (Singapore), pp. 122—32.

Government of the Republic of Indonesia (1972), 'Petroleum in Indonesia', in *U.N. ECAFE*, Vol. 1, pp. 144—28.

Graves, R. R. & Weegar, A. A. (1973), 'Geology of the Arun gasfield', *Proceedings of the 2nd Annual Convention, Indonesian Petroleum Association*, pp. 23—51.

Gribi, E. A. (1973), 'Tectonics and oil prospects of the Moluccas, eastern Indonesia', *Geological Society of Malaysia, Bulletin 6*, pp. 11—16.

Grossling, B. (1976), *Window on Oil* (*The Financial Times*, London).

Gwinn, J. W., *et al.* (1974), 'Geology of the Badak field, East Kalimantan', *Proceedings of the 3rd Annual Convention, Indonesian Petroleum Association*, pp. 311—31.

Hadikusumo, Djayadi, *et al.* (1976), 'Possible energy sources in Indonesian volcanic areas', in Halbouty, *et al.* (1976), pp. 135—9.

Haile, N. S. (1970), 'Notes on the geology of the Tambelan, Anambas and Bungusan islands, Sunda Shelf, Indonesia', *C.C.O.P. Technical Bulletin, No. 3* (Bangkok), pp. 55—75.

Halbouty, M. T. (1970), *Geology of Giant Petroleum Fields*, American Association of Petroleum Geologists (Tulsa).

——, *et al.* (1976), *Circum-Pacific Energy and Mineral Resources*, American Association of Petroleum Geologists (Tulsa, Oklahoma).

—— & Moody, J. D. (1979), 'World ultimate reserves of crude oil', paper presented at *World Petroleum Conference* (Budapest) (unpublished).

HAMILTON, W. (1973), 'Tectonics of the Indonesian region', *C.C.O.P. Technical Bulletin, No. 6* (Bangkok), pp. 3–10.

HARIADI, N. (1980), 'Kemungkinan, minyak dan gas bumi di Indonesia', *Seminar Sumber Daya Energi di Indonesia Ikatan Ahli Geologi* (Bandung) (unpublished).

HARTONO, E. E. & MANALU, P. (1974), 'Petroleum exploration activities in Indonesian offshore areas, January 1973–May 1974', *C.C.O.P., Proceedings of the 11th Session* (Seoul), pp. 215–22.

HASAN, MADJEDI, *et al.* (1977–8), 'The discovery and development of Minas field', *Proceedings of the Southeast Asia Petroleum Exploration Society, Vol. 4*, pp. 138–57.

HATLEY, A. G. (1974), 'Oil and gas exploration and development programs—the prerequisites', *Petroleum News*, September, pp. 25–7, 57.

——(1976), Offshore Petroleum Exploration in East Asia, an Overview, *Offshore Southeast Asia Conference, SEAPEX Program* (Singapore) (unpublished).

——(1978), 'Asia's oil prospects and problems: an overview of petroleum exploration activity in East Asia', *Offshore Southeast Asia Conference, SEAPEX Program* (Singapore) (unpublished).

HEPPLE, P. (ed.) (1966), *Petroleum Supply and Demand* (London).

——(ed.) (1973), *Outlook for Natural Gas—A Quality Fuel* (Essex).

HERFINDAHL, O. C. (1969), *Natural Resource Information for Economic Development* (Baltimore).

HOBSON, G. D. (1954), *Some Fundamentals of Petroleum Geology* (London).

—— & TIRATSOO, E. N. (1975), *Introduction to Petroleum Geology* (Beaconsfield, England).

HOPPER, R. H. (1976), 'The discovery of Indonesia's Minas oilfield', *Petroleum News*, June, pp. 12–18.

HUFFINGTON, R. M. (1979), 'Growth potential of Indonesia's petroleum industry', *Proceedings of the 8th Annual Convention, Indonesian Petroleum Association*, pp. 19–22.

HUNTER, A. (1965), 'The oil industry: the 1963 agreements and after', *Bulletin of Indonesian Economic Studies*, September, pp. 16–32.

_____(1966), 'The Indonesian oil industry', *Australian Economic Papers*, June, pp. 59–106.

_____(1971), 'Oil developments', *Bulletin of Indonesia Economic Studies*, March, pp. 96–113.

ICHORD, R. F., JR. (1976), *Energy Policies of the World: Indonesia* (Center for the Study of Marine Policy, University of Delaware).

Indonesian Perspectives (1974), December.

'Indonesia's Pertamina', *Petroleum Economist* (1975), March, pp. 99–101.

ION, D. C. (1975), *Availability of World Energy Resources* (London).

_____(1980), 'Classification of additional resources', *Petroleum Economist*, January, pp. 27–8.

_____& JAMIESON, W. (1966), 'Reserves in relation to demand' in Hepple (ed.), *Petroleum Supply and Demand* (London), pp. 1–22.

I.P.A. Professional Division (1979), 'An analysis of activity in the Indonesian oil industry 1972 through 1978', *Proceedings of the 8th Annual Convention, Indonesian Petroleum Association*, pp. 71–88.

_____(1980), 'Indonesian petroleum industry during the production-sharing era', *Proceedings of the 9th Annual Convention, Indonesian Petroleum Association*, May (unpublished).

IVANHOE, L. F. (1979), 'Petroleum prospects of non-OPEC LDCs', *Oil & Gas Journal*, 27 August, pp. 144–51.

JANABI, ADNAN AL–(1977), 'OPEC reserves, production and exports: a long range view', *OPEC Review*, April, pp. 13–19.

_____(1979a), 'Production and depletion policies in OPEC', *OPEC Review*, March, pp. 34–44.

_____(1979b), 'The concept of conservation in OPEC member countries', *OPEC Review*, Autumn, pp. 16–26.

JEFFRIES, K. G. (1980), 'The Sanga-Sanga field', *9th Annual Convention, Indonesian Petroleum Association* (unpublished).

JOESOEF, A. K. (1979), 'Prospect of the development of a petrochemical industry in Indonesia', *Proceedings of the 8th Annual Convention, Indonesian Petroleum Association*, pp. 89–104.

JOHNSON, M. (1977), 'Recent developments', *Bulletin of Indonesian Economic Studies*, November, pp. 34–48.

KATILI, J. A. (1973), 'Plate tectonics and its significance in the search for mineral deposits in western Indonesia', *C.C.O.P.*

Technical Bulletin, No. 7 (Bangkok), pp. 23–38.

KEHRER, P. (1978), 'World energy resources—limits from today's geological standpoint', *National Resources Forum*, January, pp. 157–69.

KENYON, C. S. (1977), 'Distribution and morphology of early Miocene reefs, East Java Sea', *Proceedings of the 6th Annual Convention, Indonesian Petroleum Association*, pp. 215–38.

KING, R. E. (1971), 'The East Asia shelves—a new exploration region with high potential', *C.C.O.P. Technical Bulletin, No. 4* (Bangkok), pp. 153–63.

KINGSTON, J. (1978), 'Oil and gas generation, migration and accumulation in the North Sumatra basin', *Proceedings of the 7th Annual Convention, Indonesian Petroleum Association*, pp. 75–104.

KLEMME, H. D. (1975), 'Giant oilfields related to their geological setting: a possible guide to exploration', *Bulletin, Canadian Petroleum Geologists*, Vol. 23, 1, pp. 30–66.

KNOWLES, R. S. (1973), *Indonesia Today* (Los Angeles).

KOESOEMADINATA, R. P. (1978), 'Sedimentary framework of tertiary coal basins of Indonesia', *Proceedings of the 3rd Regional Conference on Geology and Mineral Resources of Southeast Asia* (Bangkok), pp. 621–39.

____& NELSON, V. E. (1970), 'Mineral resources in Indonesian development', in H. W. Beers (ed.), *Indonesia: Resources and their Technological Development* (University Press of Kentucky, Lexington), pp. 117–39.

KRAL, P. (1978), 'Energy planning in developing countries', *Natural Resources Forum*, July, pp. 379–83.

KUSUMAATMADJA, MOCHTAR (1974), 'Mineral resources exploration and exploitation and the law in the Southeast Asian region: Indonesia', *Ekonomi dan Keuangan Indonesia*, March, pp. 3–41.

LEUCH, H. LE & MASSERON, J. (1973), 'Economic aspects of offshore hydrocarbon exploration and production', *Ocean Management*, December, pp. 287–325.

McCAWLEY, P. (1978), 'Some consequences of the Pertamina crisis in Indonesia', *Journal of Southeast Asian Studies*, March, pp. 1–27.

MAGGERT, K. W. & ANWAR, A. S. M. (1976), 'Use of a total field numerical simulator to plan Badak field gas government',

Proceedings of the 5th Annual Convention, Indonesian Petroleum Association, pp. 243–59.

MAGNIER, Ph. & SAMSU, BEN (1975), 'The Handil oil field in East Kalimantan', *Proceedings of the 4th Annual Convention, Indonesian Petroleum Association*, pp. 41–61.

MAINGUY, M. (1970), 'Regional geology and petroleum prospects on the marine shelves of eastern Asia', *C.C.O.P. Technical Bulletin, No. 3* (Bangkok), pp. 91–107.

——(1976), 'Economic problems related to oil and gas exploration', *Institute of Southeast Asian Studies* (Singapore).

MATHAREL, M. DE, *et al.* (1976), 'Case history of the Bekapai field', *Proceedings of the 5th Annual Convention, Indonesian Petroleum Association*, pp. 69–94.

MATTES, E. M. (1979), 'Udang field: a new Indonesian development', *Proceedings of the 8th Annual Convention, Indonesian Petroleum Association*, pp. 177–84.

MATTHEWS, A. G. (1979), 'Review of pattern stream injection in Duri field, Sumatra', *Proceedings of the 8th Annual Convention, Indonesian Petroleum Association*, pp. 479–94.

MERTOSONO, S. & NAYOAN, G. A. S. (1974), 'The tertiary basinal area of Central Sumatra', *Proceedings of the 3rd Annual Convention, Indonesian Petroleum Association*, pp. 63–76.

MEURS, A. P. H. VAN (1971), *Petroleum Economics and Offshore Mining Legislation* (Amsterdam).

MEYER, R. F. (1977), *The Future Supply of Nature-made Petroleum and Gas* (New York).

—— & HOCOTT, C. R. (1977), 'Summary', in Meyer (ed.), pp. 1–21.

MILTON, A. C. & SUMANTRI, B. (1978), 'The possibilities of marginal field development in Indonesia', *Proceedings of the 7th Annual Convention, Indonesian Petroleum Association*, pp. 433–48.

Ministry of Mines (1969), *Minerals and Mining in Indonesia* (Jakarta).

MOODY, J. D. & GEIGER, R. E. (1975), 'Petroleum resources: how much oil and where?', *Technology Review*, March/April, 1975, pp. 39–45.

MOTAMEN, H. (1979), 'Economic policy and exhaustible resources', *OPEC Review*, March, pp. 45–75.

MULHADIONO, *et al.* (1978), 'The middle Baong sandstone unit as one

of the most prospective units in the Aru area, North Sumatra',
Proceedings of the 7th Annual Convention, Indonesian Petroleum Association,
pp. 107–32.

NAYOAN, G. A. S., SUARDY, ATIK & SUYANTO, F. X. (1979), 'Nota
tentang cadangan minyak dan gas bumi Indonesia: suatu
pemikiran ahli geologi', paper presented at *Annual Scientific Meeting,
Indonesian Association of Petroleum Geologists*, Jakarta (unpublished).

NEDOM, H. A. & RAMSAY, H. J., JR. (1972), 'Exploration and
development of a new petroleum province, Java Sea, Indonesia',
Proceedings of the 1st Annual Convention, Indonesian Petroleum Association,
pp. 111–24.

NG SHUI MENG (1974), *The Oil System in Southeast Asia* (Institute of
Southeast Asian Studies, Field Report No. 8, Singapore).

NOTODIHARDJO (1975), 'Hydropower resources of Indonesia', *13th
Pacific Science Congress*, Vol. 1, Abstracts of Papers (Vancouver), pp.
157–8.

ODA, SHIGERN (1973), 'The delimitation of the continental shelf in
Southeast Asia and the Far East', *Ocean Management*, September
pp. 327–46.

OPEC, *Annual Review and Record* (various years).

——(1969), *International Oil and the Energy Policies of the Producing and
Consuming Countries* (Vienna).

PARKE, M. L., JR., *et al.* (1971), 'Structural framework of the
continental margin of the South China Sea', *C.C.O.P. Technical
Bulletin, No. 4* (Bangkok), pp. 153–63.

PENGAM, N. (1978), 'Pertamina: the long march back', *Petroleum News*,
December, pp. 18–19.

PEDRICO, E. (1978), 'The outlook for Indonesia's commercial energy
supply', *OPEC Review*, December, pp. 80–99.

Pertamina (1970), *The Story of the Oil Industry in Indonesia* (Jakarta).

Pertamina Reference Book, 1974 (Singapore).

Pertamina (Geological Dept.) (1974), 'Tertiary basins in Indonesia',
C.C.O.P. Technical Bulletin, No. 8 (Bangkok), pp. 71–2.

Pertamina (1979), *Pertamina Today* (Jakarta).

'Pertamina faces the long climb', *OPEC Bulletin*, 26 February 1979,
pp. 5–6.

'Plans for Indonesian gas', *Petroleum Economist*, July 1975, pp. 247–9.

PRATT, W. E. & GOOD, D. (eds.) (1950), *World Geography of Petroleum* (American Geographical Society, New York).

PRESCOTT, J. R. V. (1975), *The Political Geography of the Oceans* (London).

'Profile of Indonesia's petroleum industry', *Bulletin of the Indonesian Business Association of Singapore*, November 1979.

P. T. STANVAC INDONESIA (1973), 'A pioneering effort in pressure maintenance—South Sumatra's Talang Akar–Pendopo oilfield', in *U.N. ECAFE* (1973), pp. 160–8.

PULUNGGONO, A. (1976), 'Recent knowledge of hydrocarbon potentials in sedimentary basins of Indonesia', in Halbouty (1976), pp. 239–49.

____(1979), 'Perspectives within, the petroleum potential in S.E. Asia: an explorationist's view', *Proceedings of the 8th Annual Convention, Indonesian Petroleum Association*, pp. 49–69.

RAY, G. G. (1974), 'Development and completion practices, offshore northwest Java, Indonesia', *Proceedings of the 3rd Annual Convention, Indonesian Petroleum Association*, pp. 201–24.

REDMOND, J. L. & KOESOEMADINATA, R. P. (1976), 'Walio oil field and the Miocene carbonates of Salawati Basin, Irian Jaya, Indonesia', *Proceedings of the 5th Annual Convention, Indonesian Petroleum Association*, pp. 41–57.

REPUBLIC OF INDONESIA (1970), 'The status of offshore petroleum exploration in Indonesia as of February 1970 and a brief analysis of potential areas', *C.C.O.P. Report of the 7th Session* (Bangkok), pp. 91–8.

____(1971), 'The status of offshore petroleum exploration in Indonesia as of February 1970 and a brief analysis of potential areas', *C.C.O.P. Report of the 7th Session* (Bangkok), pp. 91–8.

ROSE, R. & HARTONO, P. (1978), 'Geological evolution of the Kutei–Melawi basin, Kalimantan, Indonesia', *Proceedings of the 7th Annual Convention, Indonesian Petroleum Association*, pp. 225–51.

ROSENDALE, P. (1980), 'Survey of recent developments', *Bulletin of Indonesian Economic Studies*, March.

ROWLEY, K. G. (1973), 'Rehabilitation and development, Tarakan island', *Proceedings of the 2nd Annual Convention, Indonesian Petroleum Association*, pp. 217–22.

Royal Dutch Petroleum Company (1950), *The Royal Dutch Petroleum Company* (The Hague).

SAAD, JAFAR MANSUR (1979–80), 'Conservation: towards a comprehensive strategy', *OPEC Review*, winter 1979/spring 1980, pp. 153–7.

SAFER, A. E. (1978), 'International oil revisited: could the experts be wrong?', *Natural Resources Forum*, July, pp. 327–35.

SAMADIKUN, S. (1980), 'The Indonesian energy policy', *9th Annual Convention, Indonesian Petroleum Association* (unpublished).

SAMUEL, LUKI (1980), 'Relation of depth to hydrocarbon distribution in Bunyu island, N.E. Kalimantan', *9th Annual Convention, Indonesian Petroleum Association*, (unpublished).

SAUPHANOR, M. & SEGUIN, M. (1980), 'Water injection in Handil field', *9th Annual Convention, Indonesian Petroleum Association* (unpublished).

SCHEIDECKER, W. R. & TAICLET, D. D. (1976), 'Ardjuna B structure—a case history', *Proceedings of the 5th Annual Convention, Indonesian Petroleum Association*, pp. 95–114.

SCHWARTZ, C. M., *et al.* (1973), 'Geology of the Attaka oilfield, East Kalimantan, Indonesia', *Proceedings of the 2nd Annual Convention, Indonesian Petroleum Association*, pp. 195–215.

SEMBODO (1973), 'Notes on formation evaluation in the Jatibarang volcanic reservoir', *Proceedings of the 2nd Annual Convention, Indonesian Petroleum Association*, pp. 131–47.

SHELDON, R. P. (1977), 'Estimates of undiscovered petroleum resources—a perspective', in Meyer (ed.), pp. 997–1023.

SIAGIAN, M. B. & STONE, V. C. (1976), 'Reservoir simulation of the Arun field', *Proceedings of the 5th Annual Convention, Indonesian Petroleum Association*, pp. 205–18.

SIDDAYAO, C. M. (1978), *The Offshore Petroleum Resources of Southeast Asia* (Oxford University Press, Kuala Lumpur).

SIMANJUNTAK, MAUGARA & WIJAYA, S. U. (1976), 'Coal resources and potential in Indonesia', in Halbouty, *et al.* (1976), pp. 84–8.

SIREGAR, M. S. & SUNARYO (1980), 'Depositional environment and hydrocarbon prospects of Tanjung formation in Barito basin, Kalimantan', *Proceedings of the 9th Annual Convention, Indonesian Petroleum Association* (unpublished).

SOEDIONO, IR. & LOUCKS, C. P. (1976?), 'Progress of oil and gas development in Indonesia', *Offshore Southeast Asia Conference* (Singapore) (unpublished).

SOEJODO, A. K. & INDRARTO, R. M. (1975), 'Handling of heavy Jatibarang crude oil production', *Proceedings of the 4th Annual Convention, Indonesian Petroleum Association*, pp. 199–210.

SOEMBARJONO (1969), 'Petroleum offshore activities and availability of natural gas in Indonesia', *C.C.O.P. Report of the 6th Session* (Bangkok), pp. 113–17.

SOEPARJADI, R. A., *et al.* (1975), 'Exploration play concepts in Indonesia', *Proceedings of the 9th World Petroleum Congress*, Vol. 7, pp. 51–64.

SOEPRAPTONO (1973), 'Review of the results of petroleum exploration in offshore areas of Indonesia, 1966–72', *C.C.O.P. Report of the 10th Session* (Bangkok), pp. 125–31.

SOETANTRI, BAMBANG, *et al.* (1973), 'The geology of the oilfields in northeast Java', *C.C.O.P. Report of the 10th Session* (Bangkok), pp. 149–75.

SIGIT, SOETARYO (1977), 'Mineral development in Indonesia: performance and prospects', Indonesian Mining Association Seminar, 14 June (unpublished).

STANGEL, P. J. (1974), *A Suggested Strategy for the Development of a Nitrogen and Phosphate Fertilizer Industry in Indonesia* (USAID/TVA, Jakarta).

STEBINGER, E. (1950), 'Petroleum in the ground', in W. E. Pratt & D. Good (eds.), *World Geography of Petroleum* (American Geographical Society, New York), pp. 3–24.

STOMMEL, H. E. & GRAUL, J. M. (1978), 'Current trends in geophysics', *Proceedings in the 7th Annual Convention, Indonesian Petroleum Association*, pp. 133–58.

SUARDY, ATIK, SUYANTO, F. X. & HARIADI (1980), 'Note on assessment of undiscovered recoverable hydrocarbon resources in Indonesia', paper presented at C.C.O.P., *Seminar on the Methodology on Assessment of Undiscovered Recoverable Hydrocarbon Resources*, 3–8 March (Kuala Lumpur) (to be published by C.C.O.P.).

SUMANTRI, BAMBANG (1974), 'Progress report on Indonesian offshore oil production, 1971–74', *C.C.O.P. Report of the 10th Session* (Bangkok), pp. 223–39.

SUPRAPTONO (1973), 'The status of petroleum exploration in the offshore areas of Indonesia', *C.C.O.P. Technical Bulletin, No. 7* (Bangkok), pp. 75–9.

SUTAN ASSIN, N. A. D. & TARUNADJAJA, A. N. S. (1972), 'Jatibarang, the discovery and development of a new oilfield', *Proceedings of the 1st Annual Convention, Indonesian Petroleum Association*, pp. 125–39.

SUTOWO, IBNU (1972), 'Prospect of oil for our national prosperity', (Pertamina Public Relations, Jakarta), pp. 17–21.

——(1974a), 'New energy realities confronting producing and consuming countries' (address given in Honolulu, 12 September 1974).

——(1974b), 'A fair trade value for our natural resources' (address given in Tokyo, 3 December 1974).

SUYITNO PATMOSUKISMO & IBRAHIM YAHYA (1974), 'The basement configuration of the Northwest Java area', *Proceedings of the 3rd Annual Convention, Indonesian Petroleum Association*, pp. 129–52.

SWEMLE, I. (1950), 'Indonesia, British Borneo, and Burma' in Pratt & Good (eds), pp. 273–300.

SYMONS, E. (1978), 'Liquefied natural gas—some myths debunked', *Natural Resources Forum*, January, pp. 171–6.

Symposium on the economics of exhaustible resources (1974), *The Review of Economic Studies* (University of Sussex).

Technical Secretariat, C.C.O.P. (1968), 'Regional geology and prospects for mineral resources on the northern part of the Sunda Shelf,' *C.C.O.P. Technical Bulletin, No. 1* (Bangkok), pp. 129–42.

——(1972), Explanatory note to accompany the map 'Tertiary basins of eastern Asia and their offshore extensions', *C.C.O.P. Technical Bulletin, No. 6* (Bangkok), pp. 197–223.

The development of the Indonesian oil industry, *Indonesia Oil & Gas*, August 1975, pp. 10–14.

TUCKER, E. S. (1980), 'Stretching the world's resources', *Petroleum Economist*, May, pp. 212–13.

U.N. Committee on Natural Resources (1975), *Estimates of reserves and resources—evaluation and forecasts* (Tokyo).

U.N. Department of Economic & Social Affairs (1976), *World Energy Supplies 1950–1974* (New York).

U.N. ECAFE (1959), *Proceedings of the Symposium on the Development of Petroleum Resources of Asia and the Far East* (Bangkok).

_____ (1959), Annex to the Proceedings (Bangkok).

_____ (1963), *Proceedings of the Second Symposium on the Development of Petroleum Resources of Asia and the Far East* (New York), Vol. 1.

_____ (1965), *Proceedings of the Seminar on the Development and Utilization of Natural Gas Resources* (New York).

_____ (1967), *Proceedings of the Third Symposium on the Development of Petroleum Resources of Asia and the Far East* (New York), Vol. 1.

_____ (1972), *Proceedings of the Fourth Symposium on the Development of Petroleum Resources of Asia and the Far East* (New York), Vol. 1.

_____ (1973), *Proceedings of the Fourth Symposium on the Development of Petroleum Resources of Asia and the Far East* (New York), Vol. 2.

_____ (1973), *Proceedings of the Fourth Symposium on the Development of Petroleum Resources of Asia and the Far East* (New York), Vol. 3.

U.N. Economic and Social Council (1975), Committee on Natural Resources, *Problems of Availability and Supply of Natural Resources* (4th Session, Tokyo, 24 March–4 April).

U.S. General Accounting Office (1979), *Energy's Role in U.S. and Indonesian Relations* (13 April, 43 pp.; mimeo.).

VINCELETTE, R. R. (1973), 'Reef exploration in Irian Jaya, Indonesia', *Proceedings of the 2nd Annual Convention, Indonesian Petroleum Association*, pp. 243–77.

_____ & SOEPARJADI, R. A. (1976), 'Oil-bearing reefs in Salawati Basin of Irian Jaya, Indonesia', *American Association of Petroleum Geologists*, pp. 1448–62.

WEEDA, J. (1958a), 'Oil basin of East Borneo' in Weeks (ed.), pp. 1337–46.

_____ (1958b), 'Oil basin of East Java' in Weeks (ed.), pp. 1359–64.

WEEKS, L. G. (ed.) (1958), *The Habitat of Oil* (Tulsa, U.S.A.).

WENNEKERS, J. H. L. (1958), 'South Sumatra basinal area' in Weeks (ed.), pp. 1347–58.

WESTERVELD, J. (1952), 'Phases of mountain building and mineral provinces in the East Indies', *International Geological Congress, Report of the 8th Session, Great Britain, 1948*, Part 13.

WHITE, D. A. & GEHMAN, H. M. (1979), 'Methods of estimating oil and gas resources', *American Association of Petroleum Geologists Bulletin*, December, pp. 2183–92.

WHITE, J. M. & WING, R. S. (1978), 'Structural development of the South China Sea with particular reference to Indonesia', *Proceedings of the 7th Annual Convention, Indonesian Petroleum Association*, pp. 159–77.

WIJARSO (1977), 'The energy game: an Indonesian version', *Proceedings of the 6th Annual Convention, Indonesian Petroleum Association*, pp. 47–53.

——(1977), 'A doomsday scenario—and alternatives', *Bulletin of Indonesian Economic Studies*, November, pp. 49–56.

WISNOEMOERTI, NEOGROHO (1980), 'Major interests of Indonesia in the law of the sea', *9th Annual Convention of the Indonesian Petroleum Institute* (unpublished).

WONGSOSANTIKO, ABIRATUO (1976), 'Lower Miocene Duri Formation sands, Central Sumatra Basin', *Proceedings of the 5th Annual Convention, Indonesian Petroleum Association*, pp. 133–50.

WORLD BANK (1981), *World Development Report, 1981* (New York).

World Energy Conference (1974), *Survey of Energy Resources* (London).

ZAHAR D. (1979), 'The production-sharing contract: current status', *Proceedings of the 8th Annual Convention, Indonesian Petroleum Association*, pp. 39–48.

ZEN, M. T. (1977), 'The role of energy in the national development of Indonesia', *Proceedings of the 6th Annual Convention, Indonesian Petroleum Association*, pp. 21–45.

ZIARA, B. A. & FELKOVICH, M. J. (1978), 'Well testing and analysis in Indonesian reef reservoirs', *Proceedings of the 7th Annual Convention, Indonesian Petroleum Association*, pp. 327–61.

Index